Foraging in 2021 AND Edible Wild Plants Recipes:

Foraging Guide With Over 101 Edible Wild Plant Recipes On A Budget

(2 Books In 1)

Joseph Erickson

© Copyright 2020 by Joseph Erickson. All right reserved.

The work contained herein has been produced with the intent to provide relevant knowledge and information on the topic on the topic described in the title for entertainment purposes only. While the author has gone to every extent to furnish up to date and true information, no claims can be made as to its accuracy or validity as the author has made no claims to be an expert on this topic. Notwithstanding, the reader is asked to do their own research and consult any subject matter experts they deem necessary to ensure the quality and accuracy of the material presented herein.

This statement is legally binding as deemed by the Committee of Publishers Association and the American Bar Association for the territory of the United States. Other jurisdictions may apply their own legal statutes. Any reproduction, transmission or copying of this material contained in this work without the express written consent of the copyright holder shall be deemed as a copyright violation as per the current legislation in force on the date of publishing and subsequent time thereafter. All additional works derived from this material may be claimed by the holder of this copyright.

The data, depictions, events, descriptions and all other information forthwith are considered to be true, fair and accurate unless the work is expressly described as a work of fiction. Regardless of the nature of this work, the Publisher is exempt from any responsibility of actions taken by the reader in conjunction with this work. The Publisher acknowledges that the reader acts of their own accord and releases the author and Publisher of any responsibility for the observance of tips, advice, counsel, strategies and techniques that may be offered in this volume.

TABLE OF CONTENTS

Foraging in 2021
The Ultimate Guide to Foraging and Preparing Edible Wild Plants With Over 50 Plant Based Recipes

INTRODUCTION 1

CHAPTER 1 *WHAT IS FORAGING?* 4

    What Does It Mean to Forage? 4

    History of Foraging 5

    Foraging Now 6

    Foraging Skills 8

    Foraging Safety 8

    Foraging Benefits 9

    Foraging Hazards 11

    Keep Your Eyes Open 13

    How to Forage Responsibly 15

    Don't Get Lost 18

    Don't Panic 18

    Figure Out If You Will Move or Stay 20

    Be Ready for Anything 21

CHAPTER 2 *WHERE TO FORAGE?* 23

    Don't Harm Yourself 26

    Contamination 27

    How to Find Wild Edibles in a City 28

    Urban Plants You Can Eat 29

    There's a Map for That 31

CHAPTER 3 *WHEN TO FORAGE?* **33**

    SPRINGTIME FORAGING 34

    SUMMERTIME FORAGING 37

    AUTUMN FORAGING 39

    WINTER FORAGING 42

    PRESERVING YOUR FOOD 45

CHAPTER 4 *TOOLS FOR FORAGING* **48**

    ETHICAL FORAGING 55

CHAPTER 5 *IDENTIFYING PLANTS* **60**

    HOW TO IDENTIFY EDIBLE PLANTS 61

    KNOW BEFORE YOU GO 63

    IN THE WOODS 64

    WHAT NOT TO EAT 64

    IDENTIFYING POISONOUS PLANTS 69

    PLANTS IN YOUR YARD 70

    YOU TOUCHED A POISONOUS PLANT, NOW WHAT? 70

CHAPTER 6 *LIST OF EDIBLE WILD PLANTS* **71**

CHAPTER 7 *LIST OF MEDICINAL WILD PLANTS* **82**

CHAPTER 8 *LIST OF POISONOUS PLANTS* **92**

CHAPTER 9 *RECIPES FOR FORAGED PLANTS* **102**

    Stinging Nettle Spanakopita *102*

    Dandelion Fritters *105*

    Seaweed Salad *106*

    Wild Mushroom Ragu *107*

    Stir-Fried Dandelion Greens *109*

Purslane Tacos 110

Zucchini and Purslane Soup..112

Purslane Salad 114

Stinging Nettle Soup...115

Creamy Nettle Soup..117

Fiddlehead Soup 119

Dandelion and Violet Lemonade 121

Ginger, Pineapple, and Purslane Smoothie 122

Purple Dead Nettle Tea 123

Violet Syrup 124

Wild Violet Vinegar 125

Dandelion Root Chai 126

Fireweed Tea 128

Hibiscus Syrup 129

Wild Ramp Pesto 130

Polish Fermented Mushrooms..132

Garlic Mustard Pesto 134

Sorrel Sauce..135

Fennel Sauerkraut 136

Country Mustard 137

Ancient Roman Mustard 138

Pickled Blueberries 139

Pickled Fiddleheads 140

Candied Angelica..141

Strawberry Dandelion Cake...142

Douglas Fir Poached Pear and Frangipane Tart...145

Wintergreen Ice Cream 148

- Mulberry Sorbet 150
- Gooseberry Sorbet 151
- Wild Cranberry Sauce 152
- Paw Paw Ice Cream 153
- Black Walnut Snowball Cookies 154

CONCLUSION 155

Edible Wild Plants
Over 111 Natural Foods and Over 22 Plant-Based Recipes On A Budget

INTRODUCTION 159

CHAPTER 1 *HISTORY OF HERBALISM* 161

- ARABS SAVE THE GREEK SCIENCES 164
- BYZANTINE EMPIRE 166
- ANCIENT GREEK MEDICINE 168
- EARLY MIDDLE AGE EUROPEAN MEDICINE 170
- ARAB INFLUENCES 173
- BAGHDAD 175
- QUICK REVIEW 178
- CENTRAL ASIA AVICENNA 179
- REAWAKENING OF EUROPE 182

CHAPTER 2 *KNOWING YOUR ENVIRONMENT* 185

- FOREST LAND 186
- CONIFEROUS FOREST 187
- MEDITERRANEAN 188
- GRASSLANDS 189
- TUNDRA 190

Alpine 190

Rainforest 191

Desert 191

Edible Plants In Northeast US 193

Edible Plants in Southwest US 197

Edible Plants in Southeast US 200

Edible Plants in Northwest US 201

Edible Plants in Midwest US 203

Edible Plants in South Central US 204

CHAPTER 3 *COMPENDIUM OF EDIBLE PLANTS*... 206

CHAPTER 4 *COMPENDIUM OF MEDICINAL PLANTS*... 244

CHAPTER 5 *COMPENDIUM OF POISONOUS PLANTS*... 253

CHAPTER 6 *THE BASICS OF FORAGING*... 261

Make Sure You Can Be on the Land 262

Know How to Identify Plants and Forage Safely 262

Remember the Four Rs 263

Know Your Protect Species 263

Don't Take the Only Plant 264

Take Only What You Need 264

Harvest Your Plants Wisely 264

CHAPTER 7 *RECIPES FOR EDIBLE WILD PLANTS*.. 266

Buffalo Milkweed Pods 266

Cattail Rice 268

Dandy Pasta 269

Garlic Mustard Stuffed Mushrooms 270

Kale, Lambs Quarters, and Cheese Manicotti 271

Purslane Egg Cups 273

Stuffed Milkweed Pods 274

Weed Burgers 275

Wild Potato Pancakes 276

Wild Roasted Cabbage 278

Buttered Chickweed 280

Plantain Salad 281

Blueberry Labrador Tea 282

Burdock Tonic Tea 283

Healthy Heart Tea 284

Highbush Cranberry Juice 285

Immune Boosting Coffee 286

Fennel and Angelica Cookies 287

Bee Balm Cookies 288

Coltsfoot Sorbet 289

Dandelion Banana Bread 290

Honey Cattail Cookies 291

Nutty Plantain Snack 292

Pine Cookies 293

Pine Rum Balls 294

CONCLUSION **295**

INTRODUCTION

First off, I would like to thank you for choosing this book, and I hope that you find it informative and helpful no matter what your goals may be. Throughout this book, we will discuss what foraging is and how you can safely forage for some of your food.

Foraging is something that most people have heard of. We learned in history class that our ancestors, the hunter-gatherers, were foragers. That was their only means of finding food. But then times changed and grocery stores became our main source of produce. Sure, they are convenient, and you can find food there that you would never find in your backyard, but they will never be as fresh as what you can forage.

A person who forages for part of their food will become more independent in their food supply. They won't be handing over as much of their money to large corporations. They also know exactly where their food came from. When shopping at the grocery store, you don't know as much about the produce. You aren't sure how long it has been since they were picked, or how long they have been sitting on the shelf. Plants are always better when eaten as close to picking as possible. They have a higher nutrient content at that point. But most of the food at the store has traveled hundreds of miles, and they were picked before they were ever fully ripe. Plus, if you want to make sure you aren't consuming any chemicals, you are going to have to get organic produce, which can get very pricey.

A better alternative is learning how to forage, which is what your ancestors would have done. Foraging can sound scary or even weird, especially in this day and age. That's why you have this book, though. It can help guide you through the process, so you know what to do and how to do it safely.

We will go over everything you need to know. We'll discuss some of the best places to forage for food, even if you live in an urban setting. We'll also go over what you should look for during certain times of the year, and contrary to what some people may believe, you can still forage during the winter. You'll find all of the tools that you will need for foraging as well.

Of course, we will go over how to identify plants so that you can safely forage safe foods that won't end up harming you. We even have a chapter devoted to talking about plants that are poisonous to humans so that you never eat something that you shouldn't.

The great thing is, a lot of these plants that you will be foraging for also have medicinal properties. Not only can they make a delicious dish, but they can also be turned into medical products that can help heal different ailments. Of course, you should not allow them to replace a regular doctor, only as a supplement to minor illnesses.

Foraging is a lot of work, but it is rewarding work. And you will be amazed at how many edible plants are in your backyard that you may have been mowing down for years. Adding foraging to your routine can help you save money, eat healthier, and feel better about your impact on the Earth.

As long as you refer back to this information often, and use common sense, you can safely forage for food without harming yourself or the planet.

Before we get started, I would like to ask that if you find any part of this book helpful, please leave a review. Again, thank you for choosing this book.

CHAPTER 1

What Is Foraging?

Foraging for food in the wild has become more popular in the past few years. This has become a trendy activity for some people but a total lifestyle for others due to the awareness about eating fresh, local, organic ingredients.

What Does It Mean to Forage?

Foraging normally refers to looking for, identifying, and then collecting food from the wild like shellfish, mushrooms, nuts, fruits, plants, and herbs. It is about preserving, cooking, and eating nutritious food that is healthy while understanding natural resources.

Most people have fond memories of picking blackberries with the family. This is a perfect example of foraging that many people have enjoyed at one time or another during their life. Elderflowers, sloes, and apples have also been picked in the wild to make delicious preserves in our homes.

Ireland and the United Kingdom have been blessed with a huge range of edible plants, fruits, and mushrooms that grow near the villages plus in the wild. You'll be surprised to find that a lot of the plants will grow in places you frequent, your backyard, or your local park.

Cherry trees that have been planted in parks, blackberry brambles like taking over unused lots, and hazel trees that line the street drop lots of nuts that are waiting to be picked up. By having a good knowledge of certain plants could help you find precious edibles like some wild mushrooms.

Any of these ingredients could be used to cook stews and soups or preserved as chutneys and jams. You could even make your own liquors, wines, or teas to get their medicinal properties.

History of Foraging

Foraging has been around since the beginning of the human race. The hunter-gatherers would sustain their tribes by gathering plants in their environment. They would also hunt animals. This continued until they could invent agriculture and farming as a better source of food. All the food that they ate had to be gathered by their own hands like mushrooms from trees that are decaying, wild greens from fields, and berries found on bushes. With a lot of practice, they learned how to identify the plants that are edible, poisonous, and which ones can be found at various times throughout the year.

Foraging played a huge role during hard times that were brought on by hunger and poverty. During World War II, rosehips gave soldiers vitamin C when there was a shortage of oranges, and dandelion root and acorns were used as a substitute for coffee.

Households would supplement their pantry with medicinal plants and food that were gathered in the wild until the first supermarkets were built. Even though gathering food in the wild is still part of daily life for many who live in rural areas in Europe, having a good knowledge of plants that grow in the wild has been forgotten.

Foraging Now

The food system today is a lot different. In the United States, most of our food is grown on large farms that aren't anywhere close to the people that will eat the food. For most people, "gathering food" simply means taking a trip to their local supermarket. Most probably haven't seen food growing out of the ground. Until we bring it into our houses, we don't have any connection to our food. We are limiting our choices to unhealthy foods that don't give us any valuable nutrients.

It doesn't need to be this way. In this modern world of ours, it is possible to find and pick some wild plants that aren't just edible but very nutritious. It can be a bit risky since we don't possess the knowledge of our ancestors. If a person chooses to march into the woods and starts munching on the first thing they find, they could end up facing a horrible tummy ache or worse.

It can take you years to be able to know every wild plant that could be eaten. What is more important is you have to recognize the plants that you absolutely should not eat. There are common plants that beginners can easily identify and enjoy. Adding some dandelion greens to your salad or picking some wild raspberries or blackberries while hiking will give you a chance to enjoy some fresh food that is free and delicious. It also connects you to nature and your ancestors. Plus, there is a thrill when you eat things that you have found and picked with your hands that could give you an appetite for learning more about all that nature can give us.

Naturally occurring foods are healthy was to supplement our diets. This isn't the only reason why people like to forage and take things back to the old wisdom and ways. There does seem to be an awareness about where the foods we eat actually come from.

Due to some recent scandals within the food industry, raising the awareness for leaving carbon footprints and helping the environment has caused many people to think about different ways to find our food, and this, in turn, will bring us closer to nature.

Foraging doesn't apply exclusively to the people who live in the country, but it also includes finding foods on the side of the road, in verges, and parks in a large city. Most of the time, parks will have plants that you won't ever find growing in the wild, and this gives us some very interesting finds.

Wildlife tourism has attracted a lot of people to begin foraging as a recreational hobby. This allows you to get away from our daily lives and find new and amazing ingredients that taste great and will enhance our meals.

Eventually, for some people, foraging is healthy and will help empower us. It doesn't just help us satisfy our nutritional needs but can help us understand the way nature works. Foraging can totally change how you see your surroundings and the world.

Foraging Skills

Foraging hasn't changed a lot since the beginning of humans. There are leafy plants, roots, berries, mushrooms, medicinal herbs, and nuts that are very nutritious and edible. Foraging is a skill that has been forgotten by most people. Our ancestors might have known about harvesting dandelions to make salads and wine, and they may have known that elderberries can be used to make cough syrup and wine, but most people who live in cities of all sizes have become distanced from the plants and the knowledge of these plants. Foraging skills have almost been completely lost.

Foraging Safety

While teaching a children's survival class one day, the subject of how dangerous it was to eat wild plants came up. One parent spoke up and stated that while he was an Army Ranger, he had to take many soldiers off the line because they got sick from eating something they found in the wild.

If a person is hungry enough and they see a plant full of berries, it would be natural to think that you can stay alive by eating them. If we had to choose to eat a handful of insects or a handful of greens, we are going to choose to eat the greens. The biggest problem with that is only about five to ten percent of all plants will be edible. The rest will either be unpalatable or poisonous.
This is why you have to know ways to forage safely.

You need to find a foraging guide that has good pictures and very well detailed descriptions of all the edible plants. All the descriptions need to have information about any look-alike plants that might be deadly. Elderberries are great to be used medicinally and are delicious to eat. The plant and berry can be confused with pokeberries and water hemlock. A good app, website, or book will list all the dangerous look-alikes along with the potential dangers of all edible plants.

Foraging Benefits

Finding your own food in the wild is a lot more than just being fun, even though it is that. Gathering food gives you many benefits like these:

- Connecting to Nature

Finding your own food can restore your connection to nature and the seasons that have been lost in today's world. You could watch the progression of spring by looking for the arrival of every new plant like morel mushrooms, ramps, watercress, and dandelion greens. In the summertime, you can look for ripening persimmons, wild raspberries, wild blackberries, and juneberries. When you gather these wonderfully delicious plants, you will feel like you are an actual part of nature and not just a mere observer.

- Sustainability

Any plant that you pick yourself is locally grown and organic. These plants haven't been grown with agricultural chemicals or harmful pesticides. The only water they have been exposed to is rain. They don't require fossil fuels to be harvested and transported to stores. When you go walking through the forest and gather some wild greens, they won't have any carbon footprint.

- Exercise

Foraging for food in the great outdoors will get you outside and moving. Hiking to your favorite harvesting spot, reaching and stretching to pick berries, and bending to harvest greens can be a great workout. It will be more relaxing and pleasant than spending an hour at your local gym walking on a treadmill under harsh fluorescent lights.

- Nutrition

For the most part, plants that you forage for will be more nutritious than their grocery store counterparts. The main reason why is because you get to consume them closer to harvest when they have their maximum nutrients. Wild dandelion greens will have seven times more phytonutrients than the spinach you will find in the grocery store. You can find edible crab apples that could have between two to 100 times the phytonutrients that are found in the apples at the grocery store. One good bonus is when you are outside picking your plants; you are giving your body a lot of vitamin D.

- New Flavors

Some wild foods are a bit hard to find in regular grocery stores like wild mushrooms. There are some that won't ever find in supermarkets like pawpaws. This is a fruit similar to the mango that has a custardy texture that is way too delicate to ship to stores. Foraging is usually your best chance to find these tasty and unique foods.

- Food for Free

When you forage for food, it doesn't cost anything except the time it takes you to find and pick it. Yes, your time is worth something, and it might not be worth your effort to forage foods such as potatoes. Most of the foods that can be found in the wild cost a lot of money when found in the grocery store. Pine nuts, ramps, and chanterelle mushrooms can cast more than $20 per pound, if not more.

Foraging Hazards

Even though foraging for food gives you all the benefits above, it does have some pitfalls, too, especially if the person foraging is a beginner. Some risks of foraging might include:

- Being Arrested

It is illegal for you to forage on another person's property if you don't have their permission. Many state and federal parks forbid you from gathering plants unless you are lost, and you don't have any supplies, and you are just trying to survive. In places where it is legal to forage, there are limits on the number of plants you can take. Just to make even more complicated, boundaries won't be marked clearly, so it isn't easy to know if you have wandered into a place where foraging has been banned. People who just walk into a field and begin grabbing plants might find themselves in handcuffs.

- Damage to the Environment

It isn't obvious how much of one plant you can take without killing it completely. Foragers who are overenthusiastic might strip all of the native plants out of one area. This could create an opening for invasive plants to grow and take over. Foragers could damage a delicate environment just by walking on it. They damage the topsoil, they crush plants, and they disrupt the habits of wildlife. Speaking with a seasoned forager, naturalist, or game warden can help you learn more about which areas are too vulnerable to be disturbed.

- Handling Unfamiliar Foods

Foragers who are experienced know which plants can be eaten, but they also need to know how to eat them. Most of the wild, edible plants are indigestible, bitter, or tough. If you don't know how to prepare them the right way, they aren't going to be edible. Some inexperienced foragers might find themselves looking at a basket of wild produce going to waste because they don't know how to cook with it.

- Consuming Harmful Plants

Even though most wild plants are nutritious and tasty, some might be poisonous, and most of them look like plants that are edible. Beginning foragers are most likely to mistake a plant that is poisonous for one that is edible. Christopher McCandless wrote a book on edible wild plants, but the book was published posthumously after he ate a poisonous plant in the Alaskan wilderness and died. The book is called *Back to the Wild*. Even if you are positive that the plant isn't toxic, it could make you sick if it was contaminated by chemicals, animal waste, or pesticide residue.

Keep Your Eyes Open

The biggest rule when foraging is to keep your eyes open. You might be surprised to find new foods or plants every time you look at the ground. The biggest appeal for foragers is finding, identifying, and harvesting foods that grow in the wild without any help from humans. If you are armed with some information, you can venture into a forest close to you, city part, and even a parking lot and bring dinner home. Just remember to follow all local park laws about picking plants.

- Chefs and Foraged Foods

Chefs at high-end restaurants are getting interested in foraged foods, and this has given foraged foods a boost recently. They like foraged foods because it enhances their dishes. It can also help them supplement their menus by using local ingredients that make their dishes stand out. Farmer's markets are now being filled with delicacies such as morel mushrooms and ramps along with other edible plants like lamb's quarters, nettles, and dandelion. Chefs have been working hard to get the word spread about the easiness of finding these goodies in your own backyard.

- Foraging Begins with Learning

Foraging seems like it would be an easy solution for a lot of problems with our food system, like access, nutrition, and sustainability. Eating wild is an easy way to eat sustainable if you do it responsibly. Since foraging is free, anybody is able to do it. It also provides urban dwellers access to ingredients that are more nutritious, better quality, and tastier than why they have available. Why is it so hard to get people to understand how these foods can benefit them?

One big problem is most people don't like the thoughts of eating things straight from the ground. There is an "ick factor" that some people just can get over, but with some education about the plants that are edible and the location they grow, can fix this. Most people will look at these kinds of foods as "starvation" foods. They would only eat them if it was their last resort to keep them from starving. There are many different reactions from: "I won't eat anything that has been picked from the dirt." or "this is very empowering." If these people would just stop and think about the foods that are on the grocery store shelves, they have come from the ground somewhere.

- Foraging Regulations

Not having enough information isn't just a top-down problem. Local governments still have a lot to learn when it comes to foraging being viable. This is why Berkeley Open Source Food introduced some policies to legislators on ways that foraging could supplement government assistance. One article advocated for governments to open public lands for forging and would give training to help people who live in areas where food isn't readily available, ways to identify, and then harvest the plants correctly. People look at these areas as "food deserts," but if plants grow there and they are safe to eat, this it really isn't a food desert; it is actually an information desert.

Head chef at Larder, located in Cleveland, Jeremy Umansky, is a licensed mushroom hunter who says foraging helps bring more food to his restaurant. Foraged plants make up a big percentage of his menu.

He states that anybody can forage. When you decide to process, sell, or serve the foods you find, that's when you have to have a license. This isn't saying that just because you don't have a license, you don't know what you are doing. You might have a person who never went to college, but they are just as good of a cook as the girl next door.

How to Forage Responsibly

Leda Meredith leads foraging tours in New York and wrote several books about this subject. She has created the best practices, and help come up with a bibliography for anybody who has the courage to forage.

- You Have to Know Your Plants

Beginners to foraging need to take time to become familiar with the plants. Find a guide, go on some tours, or take a class or two. You begin with, "it has this kind of leaf, and its flower is this color," and then go down the list and make sure you can check everything off of it.

Having some knowledge of your local plants will go a long way toward getting rid of any misconceptions. People are super afraid of mushrooms because there are so many that can look like edible ones but aren't. Mushrooms have to be eaten to have any negative side effects. You might touch one of the most toxic mushrooms, but if you don't eat it, it isn't going to do anything to you. On the other hand, plants have sap that could get on your skin and cause all sorts of problems, and some even have burrs and thorns. Anybody who has brushed up against stinging nettles knows exactly what I'm saying, but in all these cases, a bit of knowledge will go a long way.

Just a reminder, if you were to miss-identify a mushroom and consume it, it can end up being a deadly mistake. One good rule of thumb is if you are not 100 percent sure about the identity of a mushroom, don't eat it.

- Know the Soil

There is a misconception that urban soil has been laced with all kinds of contaminants, including high levels of PCBs, pesticides, and herbicides. This causes some foragers to worry that the food they find might harm their health. You can learn more about your local soil by taking a sample of it and getting it analyzed by your local county office. This may cost a bit of money, but it could be worth it. In one study, soil samples were taken from three different spots in the Bay Area. Some of the tested soil had high levels of cadmium and lead, but not of it had been transferred into the plants in that area.

None of the plant's tissues were accumulating any heavy metals and were completely safe to eat. The typical plant that grows wild has built up some resistance to harsh environments and are sometimes more nutritious than what is being farmed and transported to the stores. They have developed their own mechanisms to detox.

If you don't have any access to any testing equipment, you can make a phone call to your city manager and talk with the park's department. They can tell you if they use agrochemicals in your area or if it is a parcel that was reclaimed.

- Using a Newborn Stomach

When you finally identify a wild plant, for the very first time, it can be very exciting, but you have to have some restraints. If you are foraging for foods that you haven't ever eaten before, you will want to treat your body the same way that you treat a newborn. Try a few ounces of a plant and then wait for 24 hours.

Other than making sure you take precautionary measures, you should make sure you take things slow as it can have some positive environmental consequences. For example, if you make sure you don't just shove a bunch of berries in your mouth, and instead take the time to think about how you want to use and store it, you won't overharvest from a plant.

- Never Overharvest

Other than protecting your health and taking it easy on your digestive system, having some restraint is the main rule when foraging. Think about the land where you are foraging. Never take more than you will need or more than you will use.

There are thousands of medicinal or edible plants that are thought to be invasive weeds. Plants like Japanese knotweed, burdock, and dandelion, like being bullies and crowding out all the other plants around it. You can harvest them freely without damaging the ecosystem. You will actually be helping all the native plants.

In the case of the market favorite ramps, you have to forage the plant correctly and sparingly so that there s enough left for the next season. Once ramps are in season, only cut the leaves in spring and don't dig the bulbs until July. Every part of the plant will have a different season, but if you correctly treat the plant, you can enjoy it through many seasons.

Don't Get Lost

Anybody could get lost in the woods. You might be an experienced backpacker or just a day hiker, but one wrong move might send you to a place where you might be forced to rely on your own skills and very limited supplies. If you go into the woods with the correct tools, a good plan, and some know-how, you will be able to get back to civilization with poise, and it won't matter where you go.

Don't Panic

If you realize that sunset is approaching fast and your go around a clump of trees thinking that the trailhead was just on the other side of it, but you don't see it anywhere, your first reaction might be to panic. For many people, just thinking about being lost might be enough to trigger your anxiety, but the main thing is you have to stay calm. When you made a decision out of fear, these won't ever be constructive, and any drastic measure that was taken without thinking about it could make the bad situation a lot worse.

Rather than panicking, you need to STOP:

- S: stop walking; if you continue to wander around, you might get even more lost.
- T: think objectively, so you will be able to see the situation better and avoid panic.
- O: observe your surroundings. This includes any immediate risks, all available resources, and the weather.
- P: plan to handle the situation when you have the complete picture.

Remember that not all wooded threats have been created equal. Your most dangerous one would be the lack of oxygen that could be caused by a cave-in or avalanche. The second one would be being exposed to temperatures that are warmer or colder than what your body is able to stand. Not having enough water comes in third, and the last is not having enough food. You have to look at and address every risk in the order of their severity to help increase your chances of surviving if a serious condition arises.

The main factor in deciding to stay where you are and waiting for help to arrive or how much daylight you have left depends on whether or not you are equipped to handle these threats.

Figure Out If You Will Move or Stay

In order to help you decide if you want to stay where you are or move on, you have to think about everything you know about where you are. Do you have a true sense of where you are in relation to where you should be? Do you think you might have just made a wrong turn a while back and are only a little bit off your course? If you are totally confident about where you were at about a mile back, turn back and go to that point and begin from there. If you feel lost, and you think you have been going in the wrong direction, it might be a good idea to just wait. Wandering around blindly might take you to a place where your rescuers have already looked, and this will take you even longer to get rescued. If you don't have any idea where you are, but you did tell your friend where you are going, it would make more sense to just stay where you are. You know that somebody will come looking for you.

If you think you know where you are at, but it is getting dark, but you don't have a flashlight, think about creating a shelter and begin walking when the sun comes up.

The decision of whether or not to move all depends on how knowledgeable you are about the area. If you know for certain that your car is parked on the road that runs north to south and you know that you have been heading east, you might be able to find your way back to your car. If you know a way to get to the area lake and you see a landmark you recognize in the distance, let those features guide you.

Be Ready for Anything

Any risk that you might face when you are in the wild can be reduced by being prepared, even if you don't think there is a chance of being trapped or lost. Today, we rely on battery-powered navigation and technology to help us, but there won't be a good signal in the woods. Since batteries die and cell phone signals are either non-existent or spotty, this is the last thing that you want to rely on when you're are in the woods with things that take batteries.

Begin by getting some knowledge about the region you will be in. Looking at maps could help you familiarize yourself with this area. You need to make sure you do this before you take off. When you know the place you are going better, including all landmarks and the terrain, you will not get turned around as quickly.

There are some things that you should take with you before you head out. This includes spare clothes, extra water, extra food, emergency shelter, a fire starter, a knife, water purification tablets, first aid kit, sun protection, headlamp, compass, and a map.

The extra water and food will make sure you don't go hungry if you spend a night in the woods, spare clothes are going to protect you from unexpected weather that might pop up or temperatures you weren't expecting. An emergency blanket can be used as a shelter. Headlamps are better than flashlights since they let you use your hand while building a shelter or starting a fire when it is dark. Any source of light is going to help you if you are stuck outside all night.

A whistle would be a good thing to take with you if you get lost. Its sound will travel a lot farther than your voice, and it doesn't use as much energy than screaming for help. Three blasts on the whistle is a universal signal for distress.

The most important thing that you could do before any trip is to share your plans with your family and friends. By doing this, your rescuers are going to know where they need to focus their search. Knowing that there is somebody who will miss you if you don't check-in at the right time could go a long way to reduce your panic.

CHAPTER 2

Where To Forage?

For people who live in urban communities, their local farmer's market is a great place for them to find and or sell wild foods. I love seeing wild edibles in neat little bundles on tables at the market. Paying five dollars for a bunch of ramps or $25 for a quarter of a pound of mugwort that you could find for free isn't appealing at all. How funny would it be if you get home with your bundles just to see the same thing growing just ten feet from your front door? This means you need to start looking around your neighborhood. As you learn new plants, you will start noticing it growing everywhere. Even if you live in the city, there might be places where you can find amaranth and lamb's quarters growing abundantly. You might find pounds of ripe black cherries and juneberries that you can get easily. Try to make friends with your neighbors and local community growers.

Talk with the local park caretakers and keep them from killing all the invasive edibles. You might be able to join a day of volunteering to weed your local park, and you get to take home what you dig up. If diplomacy doesn't work for you, I will go where I know it grows abundantly and forage what I want.

City planners and landscape architects are adding more edibles into their projects. If it isn't happening in your area, try to encourage them to do so. I am lucky enough to have a great local classroom, which is Brooklyn Bridge Park. Its designers and creators included many edible plants. You can't forage anything, but it is a great place to go and learn about edible plants. Plus, you will get the chance to teach others about them. You might be surprised to find some of these plants in your local nursery. This type of natural partnership needs to be more common in many areas.

If you have access to nature, you are very lucky. First of all, be respectful of nature. Forage in moderation and don't stomp that native plants. If you live in the city but want to forage for wild plants, talk with your local preserves and land trusts and make friends with some landowners who might let you forage on their land. They are out there; all you have to do is ask. If you own land, think about all the wild edibles that might be on your land and the best way to conserve and curate the land.

Even if a large part of your property is managed in ways like livestock, garden, or a lawn, there is a good chance that you have edible trees, shrubs, or weeds growing somewhere.

Some of the best forages for me have come from fencerows and farm meadows, but only do this if you have the landowner's permission. It is all about talking to the people who are in charge and asking them nicely. Normally, the answer will be an amused affirmative. Weeds in the middle of crop rows can be gathered as long as they haven't been sprayed with fertilizers or herbicides.

Growers and gardeners are very lucky. They have lots to draw on in terms of wild flavor. It all depends on your hardiness zone and microclimate, but most wild edibles can be cultivated. Bring some of your forages and plant them in your backyard.

State forests and parks are great places for you to explore. Even though state laws are going to vary, there is normally a fine for harvesting edible mushrooms and fruits for personal use. Picking mushrooms, nuts, and fruits is legal in US National forests. Harvesting fungi, plants, and other items are allowed in areas that are run by the Bureau of Land Management. Just check with your local office before you pick anything to make sure there aren't any preservation areas, protected species, or local statutes where you want to forage.

Harvesting fungi and vegetation in US National Parks are regulated. There are some parks that let you harvest specific nuts, berries, and fruits as long as you harvest them by hand, and they are used for your personal use. They could set limits to the quantity or size that is harvest. They could define a certain area where edible can be harvested and restrict the consumption and possession of wild edibles to an area of the park.

Private properties like farms, hunting grounds, and homes give you excellent foraging opportunities. You just need to make sure you ask the property owner's permission before you begin digging. It's a good idea to harvest near farms that use organic produce so that you don't ingest any dangerous chemicals.

Disturbing vegetation, fungi, and wildlife aren't allowed in nature preserves or centers. Just contact your local one to make sure.

Don't Harm Yourself

Foraging is something most people do on their own, but it does carry some big responsibilities. You have to make sure you inform yourself. Treat any and all information with an open mind and circumspection. Double-check, cross-reference, and remember that anybody will put things on the internet whether or not they actually know what they are talking about. Sometimes, published authors will get their facts wrong.

You have to be 100 percent sure you are identifying the mushroom or plant correctly before you eat it. Eat only the part of the plant that you are supposed to eat and during the correct growth stage. If the recipe you find says you should cook it, you need to cook it.

If you eat a new food you haven't eaten before, there will be a possibility you might have an allergic reaction. Try some of it, and then wait 24 hours to see if you experience any sort of reactions like vomiting or swelling.

Foraging can be done alone or with a group. Add wild foods into your menu means you have to find them, and finding them might lead to meeting with other people who have similar passions. If you are a loner who likes to do things along, then you will have a feast for a table of one.

Contamination

Even though wilderness and rural areas are thought of as pristine, looks can be very deceiving. Conventional farms, mining sites, power lines, roadsides, and railroad tracks are sources of contamination. Even areas that aren't close to human developments could be home to parasites and bacteria that is spread through animal droppings.

- Water Contamination

Foragers need to be cautious in areas that might contain animal feces. If you are harvesting plants in water, take some extra steps to prevent you from contracting giardia. This is an intestinal parasite that can cause watery diarrhea, vomiting, and cramps. Getting giardia is higher in areas where there is a lot of livestock since runoff from manure can get into the water system and can introduce the parasite. Wild animals have been known to carry this disease, too. This is how it got its nickname of "beaver fever." The good news is that giardia can be killed by cooking the foods, so it is fairly easy to stay away from that one.

- Pesticides

Herbicides are normally sprayed along power lines and railroad tracks to get rid of the overgrowth of vegetation. Conventional farms will spray pesticides on their crops to keep bugs off their crops. Some people think that ingesting edible wild plants that grow close to where pesticides have been applied is the same as eating produce from the grocery store. This is so false. The EPA is in control of the types of pesticides that can be used on food crops when they can be used, and the potential for residue. When you are foraging near conventional farms, power lines, or railroad tracks, you have no idea when pesticides were applied last. This means how much residue on the plant could be a lot higher than what you normally find in the grocery store. If it isn't a farm that produces food for humans to eat, they are spraying pesticides that are not safe to be eaten.

Pesticides, even the ones that are thought of as being safe for foods, have some risks for humans like some short-term toxicity to some long-term problems like reproductive disorders and cancer. Be leery of water around these places as they could contain pesticides from runoff. Agriculture is the main polluter of streams and rivers and the third polluter of reservoirs, ponds, and lakes.

How to Find Wild Edibles in a City

If a forager looks for wild edibles, what exactly do urban foragers do? Well, they look for plants that are edible in the city. Urban foraging is a large movement for guerrilla gardening, urban homesteading, and sustainable living. New York City, Portland, and San Francisco are leading the way by creating urban foraging classes and communities to help urban forages learn what plants are safe to eat and which ones aren't. Most of these groups will have websites where their users can share the information they found and create maps to show where they found the food. The Portland group lets browsers search for food by type of food or location and lets them add newly found items.

This is different from the growing "freegan" movement. A freegan is a person who doesn't want to participate in the consumer economy and goes "dumpster diving" to find food and other items. There might be some urban foragers who are also freegans, but most foragers don't feel comfortable digging through dumpsters for perishables no matter how safe freegans say these foods are.

Urban Plants You Can Eat

All the edible plants around you might vary a lot from what is available in your best friend's area. The most popular and easiest plants to identify, grow in various places.

- Dollarweed or Pennywort

This weed is completely edible. The round, young leaves have a crisp, pleasant taste similar to a snow pea or celery. The good news is that the plant contains a compound that has been proven to reduce blood pressure and relax blood vessels. If you are fortunate enough to live in the south, then you probably have these plants in your backyard. Rather than killing them with poison, try eating them. Since they do love damp environments, you need to wash them well before eating them.

- Rumex or Dock

Rumex or Dock is another common weed. There are several varieties that exist, and its leaves are best in early spring. It is a member of the buckwheat family, and it is related to chard. The leaves are edible, and you don't have to cook them, but there are times when they might taste very bitter. You should only eat them raw in moderation since they do contain a high concentration of oxalic acid. Oxalic acid can bind up nutrients in your food, and when you consume large amounts, it might lead to calcium and other mineral deficiencies. It would be best to cook the leaves before you eat them. Cook them just like you would spinach. The seeds of dock root are also edible, but the seeds do have a lot of chaff to them.

- Smilax

This is an easy weed to identify since it is a vine that has both tendrils and thorns. Most people think this plant is invasive, and they don't realize it is edible. Any tender part of the vine that will easily snap off with your fingers can either be cooked or eaten raw. Its taste is crispy, and light and you should treat it like asparagus. Even though you can eat it raw, consuming too much uncooked could make your stomach sour os it would be best if you steamed it and used it as a side dish. An extraction of the roots of some varieties was the original root beer or sarsaparilla. The root is starchy but a great source of calories and nutrients.

- Aloe

Most people have aloe plants for burns, but you can use the plant to make your own aloe juice. Aloe juice is very healthy because it contains natural anti-viral, anti-fungal, and anti-bacterial agents. This is why it is so great on burns.

There's a Map for That

"If you really love your peaches and want to shake your trees," there is a map that can help you find one. This goes for berries, nuts, vegetables, and lots of other edible plants, too.

Foragers Ethan Welty and Caleb Philips created an interactive map that will show you over half a million locations across the world where vegetables and fruits are free. They called their project "Falling Fruit," and it shows all kinds of trees that hang over fences, line city streets, and in public parks from New Zealand to the United Kingdom.

The maps look similar to Google Maps, but the foraging locations are shown as dots. All you have to do is zoom in and click on any of the dots. It will pop up a box that has a description of the bush or tree that you will find at that location. The description normally includes information about the best season to harvest the produce, the quality of the produce, and how much fruit the plant will yield. It also provides a link to the particular species profile on the USDA's website. Plus, all the advice about accessing this spot.

Welty, who is a geographer and photographer from Boulder, CO, compiled the locations from municipal databases, urban gardening groups, and local foraging organizations. The map can be edited by the public.

Welty considers himself a data geek, and he felt there is power in putting everything on one map. It is like having a narrow lens for the world.

With many countries boasting about having foraging destinations, it was almost impossible for them to find all the spots. They had to rely on their contributor's honesty when it came to listing trees in locations that might have been off-limits such as fenced-in parks or private property. In most cases, the contributors told foragers to ask the property owner for permission before harvesting produce. This map has over 6,700 entries so far.

The two states that they created this map to create a community for beginning foragers. There is a lot of value when you pull a carrot out of the ground or an apple off of a tree.

Using their skills to help other realize there is a fruit tree at the end of their street where they can get fresh apples, this is a simple thing they did to reconnect people with the way food works and to get them away from thinking they can only get fresh produce from the supermarket.

The map doesn't limit the entries to vegetables and fruits. It also lists dumpsters that have excess food waste, public water wells, and beehives.

The creators say they hope their maps and all of its contributors will continue to grow so large that it influences land belonging to cities and management plans.

They want to make people understand that they can forage for food in cities. They also help create a food forest like the Beacon Food Forest, among others, that can be found across the country. They want people to rethink what cities need to look like.

CHAPTER 3

When To Forage?

Many would be surprised to find foraging can be done year yard no matter where you live. If you live in places where the weather stays warm all year, then you will always have a decent supply of produce to pick. But for those who have four different seasons, what you have to forage for will change throughout the year. You also have to think about the fact that you will need to store some of what you forage to help get you through the winter when there won't be a lot of food available.

Make sure when you are foraging that you stay away from areas that have been or currently are old sawmills, gas stations, and industrial sites. Roadside soil will often be contaminated by car exhaust and residue.

Springtime Foraging

Springtime is the best time to forage for wild greens. There are a lot of different flavors, and they will make you wonder why you have ever settled for the leaves and herbs available at the grocery store. After having a long winter full of meatier and heavier meals, the body is ready to crave refreshing spring greens. If you are shopping at the store, you will find spinach, parsley, Swiss chard, arugula, bok choy, and kale. They are tasty, but they aren't really in season as their foraged counterparts.

You have lady smock, sorrel, and wild garlic, as well as jack-by-the-hedge, plantain, and purslane. You can also enjoy elder, hogweed, nettles, dock leaves, and dandelion. All of this and more is prevalent during the spring months. These are typically easy to be found, and most people refer to them as weeds, but you'll know better the next time you see them.

Wild greens tend to have a stronger flavor than greens you by in the store. They also have higher nutrient content, especially when it comes to magnesium, iron, calcium, folic acid, vitamin C, and beta-carotene. In rural areas, these wild greens make up most of the diet of those who live there. Most of these greens can be dried and saved for later in the season as well.

Let's take a look at some of the most prevalent greens to harvest during the spring and where they are commonly found.

- Nettle

Nettles are best when they picked in the spring as they are still young and delicate. They grow at river banks, in the woodlands, hedgerows, and in fields. Make sure that you have gloves when you pick nettle, though, as they do sting. In order to remove the sting, you will need to wash the leaves and then pound them in a pestle, but that is only if you want to eat them raw. Cooking them will also get rid of the sting.

- Wild Garlic

March to June is the best time to find wild garlic. It likes to grow in the woodlands and easily spotted through its pungent smell and long, broad leaves. You can eat any part of the plant, but most people want it for its leaves.

- Dandelion

This plant, which most people view as a weed when it's in their yard, is bitter. They are easily spotted from their bright yellow flowers. They grow abundantly pretty much anywhere. All parts of the plant, including the roots, can be eaten.

- Wild Sorrel

Sorrel can be found in most people's backyard. If not, they are common in woodlands, meadows, and grasslands.

- Wild Fennel

This is more commonly found in coastal areas and is easily spotted by its long stem and feathery fronds. They also have small yellow flowers that show up between April and July.

- Wild Chervil

Also known as cow parsley, wild chervil is a roadside plant. It can also be found in meadows and hedgerows. It looks very similar to hemlock, which is poisonous, so be cautious. This should be picked in winter, and early spring as the leaves are sweeter.

- Sea Purslane

This is a salt-marsh plant and is very salty. It grows near coastlines. It is great eaten raw and as a seasoning.

- Plantain

This is a plant that you can't mistake for anything else, and while you can enjoy it throughout the summer, it is more tender when picked during April and May. Just like a dandelion, it is full of vitamins K, A, and C, as well as iron and calcium. Broadleaf plantain is useful and can easily be spotted by its larger, low-growing leaves. You can dry the leaves to have them handy during the winter for some tea.

- Raspberry leaves

These are best when harvested during the spring before the flowers show up. Pinch only a few leaves from each plant, leaving the rest of them to turn into berries. This is a great plant for women because it helps with menstrual discomfort.

- Chickweed

These are one of the first greens to appear as winter makes its transition to spring. In fact, if the winters are mild where you are, it may not die back during the fall. This is great for salads. Chickweed grows in damp and cool places and typically germinates in recently disturbed soil.

- Hairy Bittercress

This plant is similar to mustard greens and is best used in the same way. Bittercress often shows up in the same areas as chickweed but will show up a little later on.

- Clover

You probably walk all over this tasty treat. All species of clover are edible. White and red clover are the most common types, but you can enjoy what grows abundantly in your hard. The younger they are, the easier they are to enjoy.

Summertime Foraging

Summer is the time of the year when plants start attracting pollinators so that they can produce fruit and so that their life cycle can continue on. This is when we have pretty flowers that can also be eaten. A lot of the plants that you harvested in the Spring may still be available at this point, but they are often a little more bitter or tough. But there are other delicious plants that you can enjoy during the summer. Be mindful of bugs and wash your plants very well.

- Lemon Balm

This plant is in the mint family, so it is easy to spot for its mint-like appearance. It has oval leaves that are hairy with a toothed edge. When crushed, they will smell distinctly like lemon. They will form white/yellow to pink flowers later in the season that bees love. The leaves on the top tend to be better for eating raw. You can take the older leaves, but make sure they are cooked or mixed with other items to mask their harshness.

- Borage

This is a hairy-spiny leaved plant that tastes a lot like cucumber. It produces five-petaled blue flowers, which you can pick by griping the black stamen to pull off the flower from the calyx. This is commonly found on waste ground and near gardens and hedges.

- Red Clover

Again, this can be found in any grassy area. When left to grow, it can reach nearly two feet tall. It flowers from May to September. It is very sweet, and the flowers should be picked when they are about a quarter of an inch in size.

- Oregano

This plant likes to grow in dry grassland, hedges, and woodland edges. It has hairy, green, oval leaves. It will flower, but it tastes better when you can harvest it before it flowers.

- Rose

All rose petals are edible but stay away from any plants that have been sprayed. Wild rose is often found in hedges and contains curved thorns. They often bloom from June to July. The field rose is similar, but is smaller and will have white flowers.

Autumn Foraging

Just like with a garden, nature will put out its biggest bounty during the fall. While greens are abundant during the spring, fall foragers can find fungi, nuts, roots, and fruits. Let's take a look at the tasty treats the fall has to offer.

- American Persimmon

Everybody is familiar with the Asian persimmon. What most people don't know is there is an American cousin that hides out in the hardwood forests of the eastern US. They are smaller but have a great taste. Make sure they are completely soft before eating.

- Pawpaw

These are the largest native fruits in North America. They are similar to cherimoya but are most commonly found in the bottomlands in the Eastern US. They are most commonly found on riverbanks in dense thickets. Pick once the skin turns yellow and the flesh is soft.

- Madrone Berries

You can find fruits in the west as well. Madrone berries are marble-sized fruits that come from iconic western trees. Make sure they are a deep red color. They are kind of dry, so they work better mixed with other things.

- Burdock

This is sometimes referred to as gobo. This is commonly grown in Asia, but it tends to be a weed in a pasture than a cultivated plant. It grows in every corner of America except for the Deep South. It has fuzzy leaves that are shaped like an elephant eat. Its taproot can be several feet in length. It is a biennial, so it only produces leaves the first year, then it will flower and set seeds during the second year. To get the best tasting burdock, harvest it at the end of its first year.

- Groundnut

This is a perennial leguminous vine that has an egg-sized tuber that you can use just like a potato. It prefers low-lying wetlands. Its vines climb over trees and shrubs in thickets from Texas to Maine and Florida to North Dakota. Harvesting their roots kills the plant, so be nice and replant a few of the tubers in the same place.

- Acorns

There is a lot of different oak trees through North America, and all of them produce edible acorns. But they are not a nut that you can just pop into your mouth off the tree. They are inedible until they have been processed a bit. To do this, place the acorns in a jar of water for a week or more, changing the water every day.

- Hazelnut

The hazelnuts bought in the store come from a cultivated European tree. But North America has its own native hazelnut, which is tasty but hasn't been developed into an agricultural crop. They will often grow on large shrubs that are found at the edge of forests and can be found in nearly every state except for the Southwest.

- Pine Nuts

These are unique to the southwest and California, where they are harvested from different pinyon pines. Pinyon pines are found at elevations over 5000 feet and should be harvested when the cones turn green to brown.

- Chanterelles

The season for wild mushrooms varies depending on where you live, but chanterelles are commonly found throughout American during the fall. They prefer cool weather and often emerge in the forest after a very good rain. They are one of the easiest mushrooms to spot. They contain ridges on the underside instead of gills like other species.

- Hen of the Woods

These are also called maitakes. You can find them throughout the states and are commonly located at the base of hardwood trees. They especially like oaks. They parasitize the tree, so they are more likely to be on dying or dead trees. They grow in huge layered clumps that are about the size of a chicken.

- Jerusalem Artichoke

Sometimes called a sunchoke, is a wild sunflower that is found in the central US. They tend to grow up to 12 feet tall. Their leaves are up to three inches wide and eight inches long, and their yellow flowers emerge around August and September. They grow nearly anywhere, but the best tuber production requires well-drained soil.

Winter Foraging

Most people think foraging will end once the weather turns cold, the leaves are gone, and most growth has come to a standstill. But what is a person supposed to do during the winter? When we are talking about winter, we are talking about the cool night in the desert or brisk winds you get in the southeast. Places that never see snow and rarely, if ever, see ice, don't really have a lot to worry about during the winter, but those who live from around North Carolina and up on the East coast, the Midwest, and the upper half of the West coast has less to work with.

I'm here to tell you, though, it is possible to find food during the winter, but let's take a look at some things that will complicate your winter foraging.

1. It's Cold

This won't only affect what you are trying to gather, but it is going to eventually affect you. The cold will also cause the ground to freeze, which is going to limit your access to tubers and roots.

2. There Could Be Snow

Snow can cover and obscure the things that you are looking for. You will have to know how to look for clues above the snow. An oak tree is a good indication that you could find acorns under the snow. Some oak trees that will hold onto a few of their leaves over the winter, so that will help.

3. It's Wet

Many people like harvesting cattails during the winter, but having to slosh through a foot or two of water and sticking your hand down into the water and mud is going to get old quick.

4. Less than 10% of Items are Available

If you are in an extreme winter climate, most things are going to be dead or not growing. Options will be limited to around 10%, depending on your location. In winter, we will often lose indicators that help us to discover food, especially leaves. However, there are still some indicators.

You need to make sure that you pay attention to the appearance and shape of the bark on trees, especially nut-bearing ones like black walnut, horse chestnut, and oak. Take the time to learn different barks and the characteristics of nut-bearing trees. A good clue would be if there is a squirrel's nest in the tree.

There are some plants that can continue photosynthesis under the snow. Scraping away the snow can reveal chickweed, wild onion, and dandelion. You can head into the water, but be careful. If you live close to the ocean, tide-pools during low-tide can help you find seaweed, kelp, and even shellfish if you eat fish. Ponds, freshwater springs, and creeks can have cattails, and crayfish and mussels if you want those as well. Make sure you are well dressed for water foraging during the winter. Remember, it's cold. You need to make sure that you dress warmly and in layers. You are going to experience varying degrees of rest and exertion, and you will need to be able to manage your perspiration.

The best foods to look for in the winter include:

- Cattails

The roots of these plants can be washed, peeled, and used similar to potatoes. You can also dry them and turn them into flower. If you decide to go looking for cattails in the winter, make sure it is the only thing you are planning on doing. Make sure you have some waterproof boots, insulated hip-waders, and some heavy-duty rubber gloves that go all the way up your arms.

- Horse Chestnuts, Acorns, and Black Walnuts

These will be on the ground at this point. Make sure that you soak them for about three days, changing the water every day, and then you can roast them, boil them, or dry them and turn them into flour.

- Rose Hips

These are typically bright red and around a quarter to a half-inch in diameter. You can make them into jelly or use them in tea.

- Mushroom

Different fungi may appear during the winter months, likely after a brief thaw. Try looking for them on rotting deadfalls. Make sure you don't pick toxic mushrooms.

- Wild Greens

While not as prevalent, there are some greens that will show up under the snow or poking through the leaves. Rinse and enjoy them.

- Wild Fruit like Crabapples and Plums

These are easy to spot as they will still be hanging on the tree. They are great as a jelly or as a juice.

Preserving Your Food

While you can find some plants to eat during the winter, the best way to make sure you have food for the winter is to act like a squirrel. Make sure that you forage enough during the year and preserve it so that you have it on hand when winter comes. Wild foods will last longer than store-bought foods. Most greens will last you about a month when kept in the fridge. Burdock flowers can last for two months, and the roots can last three to four months. There are different ways to preserve your food. Let's take a look at a few different methods.

- Blanch and Freeze

There are some wild edibles that will have to be blanched before you freeze them, so it is best to research each one individually. Others you can simply freeze as they are. Freezing is the easiest way to preserve foods. Rinse your plants in some cold water, shake off the excess, and chop them up. Add the chopped plants into an ice cube tray and then place them in the freezer. After they are frozen, place the plant-cubes into an air-tight container or freezer bag.

Another way to freeze the plants is to spread the loosely across a baking sheet and then freeze them. Once frozen, place the edibles in a freezer bag, seal, and keep it in the freezer. Now, these frozen treats won't be able to be used in salads because once they thaw, they will lose their integrity. However, they can be used in different cooking methods. Make sure you don't refreeze anything you have thawed.

- Drying

Drying is another popular method of preserving foods. If your wild edibles are clean, make sure you don't wet them. If they are not clean, briefly rinse the dirt and dust off of the foliage, shake any excess water off, and get rid of any damaged or dead foliage.

Take the stems of the plants and tie them together in a small bundle with an elastic band or string and then hang them upside down in an airy, dry place. Make sure they are not placed in direct sunlight. You want the bundles to be small and loose to allow for good air circulation. Elastic bands are best so that you can tighten up the band as the stems start to dry out and shrink. You can use paperclips to hang the bundles onto a rope or string.

The UV rays from the sun and moisture like frost and dew can end up discoloring the plants and reduce the quality of the herbs. That's what it is best if your dry your plants indoors in a closet, attic, or a small unused space. It can actually add a nice scent to a room.

If you don't want to hang your plants to dry them, you can spread them out across a clean window screen or some other type of screen. Place the screen between the backs of two chairs so that it can have has much airflow as possible. Flip the leaves over often to make sure everything dries evenly.

You can also use a conventional oven to dry your edibles. Spread the plants out into a single layer on a baking sheet and place them in the oven at the lowest temperature. Food dehydrators can do a great job, as well. Your plants are sufficiently dry when they crumble easily. After they are dry, you can separate the leaves from the stems. Keep the dried plants in mason jars with a tight lid. Keep them stored in a cool, dry place away from heat, moisture, and sunlight. When stored properly, they typically last one to two years.

CHAPTER 4

Tools For Foraging

All types of crafters have a cache of specialty tools, and foraging is no exception. The good news is, if you have a garden, flower garden, or yard of any type, you will likely already have some of these tools. Tools will make your life easier when it comes to foraging. Not all plants are made the same, so you will need slightly different tools for different types of plants. Sometimes plants just want to fight you when you try to harvest it, especially if you are trying to dig up roots during the winter. Tools make this process simpler, and it doesn't put as much stress and strain on your hands. Let's take a look at the types of tools you will want to have before you start foraging.

- Pruners

Pruners will be what you use the most often when you are gathering and processing foraged herbs. They are able to snip right through the herbaceous stem. They can also cut through roots, small branches, and twigs. You will find that you use these more than almost any other tool. If you are only able to buy one tool to get you started, pruners are the one tool that you should get.

There are different types of pruners out there. Look for high-quality pruners and make sure that their blades can be sharpened. Keep in mind the spring and blades can wear out. There is a brand called Felco that a lot of foragers rave about. They are long-lasting, and they even offer replacement springs and blades. Make sure that you keep them good and sharp because dull pruners can damage the plant.

Look for pruners that can help to reduce hand strain and fatigue. When your pruner handles are fully opened, they should not exceed the width of the of your grasp.

- Weeding Knife, Hori-Hori, or Japanese Garden Knife

This tool is a compact and heavy-duty, wildcrafting tool and is great for weeding. A hori-hori can be used to break up the soil and dig out small to medium-sized roots. The garden "knife" can cut through the majority of clay soils, and they can even be used to pry rocks up out of the ground. They can also be used to transplant and divide roots.

A good quality garden knife can last you for many years. Along with the pruner, you can purchase a holster for your knife so that you can keep it on your belt or person for easy access. The wooden-handled variety is believed to be stronger than the plastic ones, but if you tend to lose things easily, think about purchasing one that has an orange plastic handle to reduce your chance of losing it.

- Digging Fork

Digging fork, also known as a pitchfork, is the tool you will use to help you dig up most roots. The tines on the fork will loosen up the soil and help to life branching roots out of the Earth. A digging fork is a lot less damaging to roots than a spade or a shovel. The digging fork can be used in the garden to weed or loosen up the soil. Digging forks will have sturdy and square tines, unlike that of the hay or manure forks that have bendable, flat tines. You can find affordable digging forks at almost any big box store, but you do get what you pay for. The last thing you want to happen is to have your handle break after your second foraging session.

- Shovel

There is a good chance you already have a shovel hanging around your home somewhere. Having a few different types of shovels can be helpful. You will want to make sure that you have a least one long-handled shovel that has a pointed blade. Shovels will mainly be used to excavate large, tap-rooted plants like burdock. They are also helpful in heavily compacted soils.

- Compact Shovel

You may not want to lug around a full-size shovel with you, so a compact shovel is the way to go. There are a lot of underground edibles, such as burdock root, wild potato, cattail, and leeks. You can get compact shovels in various shovels. That said, if you are planning on doing a lot of foraging or are foraging bigger items, a full-sized shovel will make the work a lot easier.

- Kitchen Scissors

Having a sharp pair of kitchen scissors is a great tool to have for gathering tender-stemmed greens like cleavers, violet, or chickweed. Pruners can end up making a mess of the job as they are supposed to be used on tougher stems, and their blade reach is pretty limited.

- Pruning Knife

A pruning knife is a hook-shaped knife that will make short work of cutting stalks and vines. You can use straight edges, but a pruning knife will make things easier. They are purposely built with the hook shape to make slicing with a single stroke possible. This knife will allow you to cut low to the ground and in various terrains. It can also help protect your dedicated survival knife. If you can't find a pruning knife, a knife meant for cutting linoleum can do in a pinch.

- Pruning Saw

Having a foldable pruning saw can be a handy cutting tool for small to medium-sized tree branches and limbs. This is a good tool to have if you plan on gathering medicinal tree barks like black birch and wild cherry.

- Sharp Compact Knife

Having a simple sharp knife is great for peeling the bark off of medicinal trees. You will want to have a good quality folding knife or a compact knife that has a sheath so that you don't accidentally cut yourself.

- Assorted Baskets

You have to have something to hold your foraged foods, right? Baskets can help you out in many different ways. They are handy when you are gathering up and drying herbs, and they are pretty as well. It is helpful if you have an assortment of baskets. You can easily find baskets at thrift stores. Try to find some that have an open weave and are flattish and broad to add in ventilation for drying loose herbs.

- 5-Gallon Buckets or Tubtrugs

Buckets get used more often than you may think. They can be pulled out for large-scale harvests, such as blueberries and elderberries, and for harvesting muddy roots. Adding some water at the bottom of the bucket will help to keep leaves and stems of herbs fresh during a car ride.

You can use repurposed food-grade buckets. Both three and five-gallon sizes are great. You can also ask for some empty buckets at the food prep section or the bakery counter at your local grocery store. Five-gallon buckets can also be bought at hardware stores. Tubtrugs are pliable buckets with handles and can be quite helpful for harvesting foods. They tend to be on the expensive side, but they will last a good while.

- Gloves

Foraging can be rough on your hands, and your fingertips will love you for keeping a pair of gloves handy for any prickly situations, such as stinging nettles or a berry bramble. Plants aren't able to run away, so they have grown their own defense mechanisms that will wreak havoc on skin. A single encounter with nettles and the stinging sensation that occurs when you touch will remind you to have gloves with you at all times. Thorns are also a fun thing to run your hand across. There is an old joke that tells you how to spot the difference between a blackberry and a raspberry thorn. All you do is grab hold of the stem, pull your hand down its length, and if your flesh is still attached, you have a raspberry bush. But, if the flesh in on the thorns and your hand is torn to bits, it's a blackberry. This "joke" illustrates a good point. Carry good gloves to handle plants is worth your while. Having two different types of gloves handy is a good idea. A thin, supple pair can be helpful for delicate tasks. A thicker leather pair is better for the moments when you will need more protection.

- Heavy Duty Chopping Knife

Heavy-duty knives are necessary when you are chopping through those tough roots.

- Breathable Bags

This is an alternative to a basket. A breathable bag is another way to store what you have gathered and will allow air to reach the plants so that they don't start turning brown and decomposing. You can find these breathable bags at just about any store.

- Sturdy Vegetable Brush

You will want to have a sturdy bristled brush to scrub the soil out of the crevices and cracks of any roots you harvest.

- Loupe or Hand Lens

This isn't a must-have, but having a jeweler's loupe at a 10 to 20 times magnification can be helpful. This will allow you to look at small botanical parts and identify them more properly. They have a higher magnification than a simple magnifying lens. There are some that have LED lights attached, which is great for viewing plants.

- Water bottle or thermos

This won't help you to gather anything, but it will make sure you don't get dehydrated. No matter where you are foraging for plants, if you plan on being outside for any length of time, you need to make sure you have something to drink. A thermos is good if you are foraging in cooler weather. Some hot coffee or hot chocolate can help warm you up if you start to feel cold.

- Field Guides

One of the best ways to learn about plants is to have first-hand knowledge about it. When you have a reliable source with you, it can help to answer any questions and get rid of any uncertainty. If you can't find a guide, which most of us can't, using multiple field guides is the best way to go. There are some field guides that use pictures, and some will only use line drawings. There are a lot of them that will make the perfect plant sample and feature that when, in the real world, what you will find will be missing flowers, have fewer leaves, and things that aren't as bold. Once you have a plant tentatively identified, verify what you think it is with at least three other sources. You will want to make sure the descriptions line up across various sources and authors.

Ethical Foraging

While this may not be a tool that you can hold in your hand, it is a tool that you can have in mind to make sure that you don't cause harm when you are foraging. Knowing that lots of people want to learn how to forage makes me happy. The fact that people are willing to spend more time outdoors, interacting with nature, is a good thing. The process of identifying, harvesting, and preparing wild edibles can bring you closer to your ecosystem.

But it is a practice that can be full of abuse. Lack of education and overzealous actions can be dangerous to the person, as well as devastating to the environment. However, if you can keep just a few things in mind and follow the best practices, foraging is a rewarding and fun experience that won't harm the environment or you.

1. Know Your Environment

While people will obviously know that they can't start foraging for prickly pears in the rainy Pacific Northwest, it is important that you understand and know what is available or what could be "at-risk" in your area. Grab a field guide from a local library, online, or bookstore, and study up on your native flora. Pay close attention to the characteristics of the plant, their growing conditions, and when they fruit or bloom. The USDA plant map is a great tool to help your research if a wild edible grows in your area.

You should also seek out foraging walks provided by professional foragers and herbalists and get a spot in their next event. Local fishermen, farmers, and hunters tend to be great resources for finding abundant plant matter.

2. Have a Foraging Plan

You should never head out into the woods without having a plan. You need to have in mind the things that you are looking for and where the best places would be for you to start looking for them, then stick to that. If you happen upon something that interests you, take a note of your location, take a sample or picture, and then consult your identification materials and field guides when you are able to.

When you do this, it will likely save you a lot of trouble and danger of misidentification while also making sure that you are ethically harvesting because that plant could be in danger.

3. Harvest from "Clean" Areas

Make sure you forage in litter, spray, and pollution-free spaces. Don't harvest things off of the roadside, in industrial areas, along property lines, or in city parks. All of these areas have the potential for contaminants and pollutants. It is also best to find light traveled and untouched areas for your foraging. This will help to ensure that the harvest is safe and clean.

4. Identify

This cannot be stressed enough. You have to learn how to accurately identify herbs and mushrooms. Don't rely on a single characteristic, such as a leaf or bloom, to ID the plant. You need to use three or more points of identification. Consider the color, bloom, leaf, fruit, stem, bark, fragrance, branches, life cycle, location, spore print, and/or soil condition.

5. Be Conservative

The Majority of ethical foragers recommend that you only harvest around a tenth to a third of a certain patch of what you see, and never from one patch. For example, if you pass by a small patch of mugwort and see that it is sparse, small, and the only spot of mugwort in the area, don't harvest from it.

You should also look at the life cycle of the plant. If you take all of the white blossoms off of the elder tree during the spring, then there won't be any berries come fall. You should only harvest what you actually need. Exercise restraint even if it is hard.

6. Leave It As Good As or Better Than When You Found It

There is nothing more frustrating than seeing your favorite spots ransacked, spoiled, and pillaged by less appreciative people. Get rid of all of your garbage, and think about having another bag with you so that you can clean up any litter or messes that you come across.

Don't drastically change the landscape for your own means. You should not chop down limbs or tress. Don't pull a tree across a stream to make a bridge. Don't start driving off-road. Don't disturb animal homes, like dens or nests. Just don't do anything inappropriate.

You should report unsafe conditions to the correct authorities.

7. Prepare and Inform

Get the equipment and clothing ready that you are going to need for your foraging. Make sure you always let another person know where you are getting ready to go, and how long you plan on being gone.

8. Check Out The Legalities in Your Area

To be an ethical forager, you also have to be a legal one. Check out any regulations and laws in your area about where it is legal to gather mushrooms and plants, and if you are going to need a permit. Make sure you never trespass. Make sure that you have permission before you start foraging on private property.

If you are certain about the legal areas to forage, this is where you may want to talk to a hunger friend as they probably know the answer to your question. They may also be able to put you in touch with somebody who will let you forage on their property. You should also be aware of the hunting season schedule and take safety measures during those times.

By keeping these things in mind, it will help you to have fun, while also being safe and legal. Restraint, proper planning, and education are key characteristics in the ethical forager. Foraging helps to fulfill some basic and primitive urges, provide you with exercise, and adds activity to a normally inactive time. It helps to bring you closer to nature and strengthens your appreciation for the world around you.

CHAPTER 5

Identifying Plants

During the late summer and early fall months, the harvest is at its peak. Driving through the country, you will see roadside produce stands along the side of the roads. You can find all sorts of yummy foods like pumpkins, apples, peaches, berries, etc. When you forage for food, it lets you feel all the joy of eating local foods without having to wait for the farmers to harvest the food for you.

Foraging, as you have probably realized, is a lot harder than just walking into the woods and picking plants to eat. You absolutely can't do this because there are too many plants out there that are poisonous and could kill you.

Don't let this fact scare you. There are many benefits to eating plants found in the wild. You just have to know how to eat wild plants safely. Once you know this, you might just save yourself some money on your grocery bill. If you have a dog that can sniff out truffles or find a patch of rare berries, you are in business.

I like to forage because I get to be outside, get my hands in the dirt, and it helps me clear my head. I love to eat the clean foods that are free of pesticides while learning about the plants that are around me. But my most favorite part of foraging is getting to eat the food I find.

I love to target invasive plants and eating them. Some of the more edible invasive plants include burdock, lamb's quarters, and garlic mustard. Since man caused the problem with invasive species, it is about time we undo what we started.

How to Identify Edible Plants

It isn't a secret that everyone loves using apps to identify plants. If you are eating plants, you need to be 100 percent sure that the plant is safe for you. PlantSnap is a great app that will help you learn how to identify plants, but let's go one step farther to keep everyone safe.

You can begin with some of the most familiar plants, such as blackberries. Even though there are hundreds of different species and they are extremely hard to tell them apart, they are all edible.

You can use your apps, but it would help you if you can gain some knowledge of some basic botany skills. If PlantSnap gives you a few suggestions about some edible plants where your life, you can narrow your search down from toxic to tasty by noticing:

- General Notes: Watch out for the plant's shape and height.
- Stalk: Some plants might have spines, spots, or other markings on the trunk or stalk.
- Flowers, Fruits, Seeds, and Cones: Pay more attention to the plant's flowers, fruits, seeds, and cones as these are their reproductive parts and are normally easy to identify the plant by.
- Arrangement of Leaves: This can get confusing very quickly, but try to notice how the leaves are arranged with as much detail as you possibly can.
- Shape of Leaves: Do the leaves look like needles, are scaly and flat, or are they actually leaves? What does their outline look like?
- Micro-Ecosystem: The area where you found the plant, is it shady or sunny? What kind of soil was it in? Does this area get lots of rain? Can you tell if the water drains away from or towards the plant? What is the altitude of where you are at? Are you in a wetland, savannah, forest, or someplace else?
- Global Location: You shouldn't waste your time looking at a plant that is native to China if you are located in England, well, unless it is an invasive species that was brought from China.

If you can learn these particular traits, it can help you identify the right plants to forage.

Know Before You Go

Before you even think about heading into the woods to find salad fixings and berries, you have to know some of the main principles for foraging. Here is a quick overview of some things you need to know:

- Find invasive species if you can.
- Know ways to prepare the food when you find it. Some plants are only safe to eat after they have been cooked. Some edible plants will have toxic parts to them.
- Know the laws in your area. You can't forage for food in National Parks or private property.
- Leave some food for wild animals. Don't be that jackass who takes every single blackberry off of the bush. There are wild animals that depend on these plants, too.
- Know all the lookalike plants. Make sure you know of all the plants that may resemble the plant you are looking for.
- Know the food you want to find. If you know a little bit about what you are looking for, getting a proper identification will be a lot easier.
- Know the ecosystem. Try your best not to put any stress on delicate ecosystems.

Now that you have a fairly good idea about how to identify plants and why you should forage, it is time to get out there and see what you can find.

In the Woods

You have realized you are lost in the woods, you are days away from civilization, and you don't have many snacks left that you packed. How did you stray off the path? You are going to need to find food very soon.

You aren't going to have to look too far. Just look around you, there is a nutritious buffet waiting to be uprooted, cracked, and plucked. You have all sorts of edible plants just waiting on you.

In spite of what most people believe, most of the plants found in the forests in North America are safe to eat. The hard part is finding the plants that are tasty and nutritious when not cooked.

Most plants in the environment will be edible. They might not taste too good or give you many calories. Basically, you are going to need to be picky if you want a very decent meal.

Here are some tips on how to find plants that will satisfy your hunger while tasting good and ways you can stay away from their toxic cousins:

What Not to Eat

Plants are very tricky. Most are edible, but one mouthful of a bad plant could be deadly. That might be a bit of an exaggeration, but there are plants that you have to stay away from all the time.

Try not to get fooled by plants that look like edibles. Many wild plants look exactly like Italian parsley, such as hemlock. This plant is what killed Socrates, the Greek philosopher.

You don't have to know one poisonous plant; you just need to know what you are eating. Basically, this is saying you should only eat a plant if you can easily recognize and know that it is safe.

The plants listed below are the most common plants found all over the world, and you might be able to find them in your backyard. Being able to identify wild plants need to be at the top of your survival skills list.

- Ground Nut

Groundnuts are a member of the pea family, and they help fix the nitrogen levels in the soil. They aren't on the most popular food list because they do have a two-year cycle. Groundnuts like moist sandy soil near the banks of a river. They spread rapidly and can be found all across the United States. Their green parts look a lot like wisteria. Groundnut leaves will be pinnate and have between five and seven leaflets with smooth edges and hairless. The flowers have a musky fragrance. Groundnuts have an edible tuber that is made up of about 20 percent protein. The tubers will be sweeter in the fall, but you can harvest them at any time of the year. Trace the fragile stem to the ground and dig down about two inches. Then gently pull to unearth the tubers. Because their skins are thin, you don't have to peel them. Never eat them raw as they can cause gas and are very sticky. Cut them into small pieces and steam for 20 minutes. Check for doneness as you would a potato. You can save the stock for another soup if you would like.

- Wild Onion

This is probably the easiest food to identify. It is going to look just like an onion. The tops will be thinner and curly. You need to double-check this food's identity by smelling it. Many wild onions will grow in forests all across the United States, and they are a good source of food. If it doesn't smell like an onion, you shouldn't eat it. Your nose is very handy for finding poisonous plants. Stay away from any plant that smells like almonds, as this could be a sign of cyanide.

- Wood Sorrel

Most people are familiar with this plant as it can be found everywhere on the Earth except at the poles. There are over 800 species of this perennial, and it can grow between six and eight inches high. It has three leaves on each stem and looks very similar to clover. The wood sorrel will have a tart flavor. I remember eating this as a child in my backyard.

- Pony Foot

This plant looks like a pony's foot. It grows in wet, swampy areas. It can be easily found in most lawns. It doesn't have a strong flavor and is great to add to salads.

- Dollar Weed

This is a very common plant that most people don't want in their yards. It has a fresh taste similar to a mix of celery and carrots and can be put into any stock. It is a member of the carrot family, and the leaves are what you will eat. The roots and stems are too hard. It has bright green leaves that are rounded, and the edges are a bit wavy. It will produce small, white flowers in July and August.

- False Hawksbeard

This plant has a crinkly, veiny, edged leaves that are slightly curled. The plant comes up in early spring. In Florida, it will grow in the shade during hotter months. It does look like a dandelion because its leaves grow in rosettes, and it has yellow flowers. Hawksbeard is different from dandelion because their stem will contain multiple stalks, and each stalk will have multiple blooms. The young leaves can be eaten as a salad green. Older leaves can be added into soups and stews as an herb. You can find this plant from Pennsylvanian down to Florida and then west to Texas.

- Bacopa

This plant can be found in any semi-wet place in the world. It is a good health food that can affect neural development and regeneration. This can help with retaining memories. It has thick, small, succulent leaves that creep along the ground at about six inches high. The leaves will be rough when you touch them, and they smell citrusy like a lemon or lime. You can add them to some hot water and have a refreshing tea.

- Tree Nuts

What plants do you need to look for? You are surrounded by leaves, but it will take a wheelbarrow full to make you feel full. If your main goal is just survival, you need to find the fattier, calorie-dense parts of the plant. Tree nuts are a great option, and they are readily available in most woodlands.

If you live east of the Great Plains, look for hickory nuts. These are considered as the most calorie-dense food found in the wild. These are produced by deciduous, tall hardwoods. The nuts are very hard to crack. They will have an outer and inner shell. It will be worth your effort if you can get into them. They do taste a lot like pecans, and the pecan you find at your local grocery store actually comes from a southern species of hickory. You don't have to soak or cook them before you eat them. Just make sure the nut is veiny, just like a pecan. Buckeyes look a lot like hickory nuts, but they are extremely poisonous. The meat of the buckeye will be rounded and smooth.

If you find yourself in the American southwest forests, try to find pine nuts from the pinyon pine. This is a scrubby evergreen in the desert. It is a great option when you are hungry. These nuts are found inside pine cones. They are easy to harvest and taste like buttered kernels. The pine nuts you find in the supermarket are imported. Chefs and indigenous Americans have been using pinyon nuts for many, many years.

Don't forget about acorns. They are edible and were a food source for the Native Americans. You do need to prepare them first. You will need to take a rock to get the nut out of the shell. If you don't have a pot, you can use a clean sock to submerge the nut's meat into some water for a few days. The water will get rid of the tannic acid inside the nut. Too much tannic acid can cause stomach problems. If you do have a pot and are soaking the nuts in the pot, you will need to change the water a couple of times.

- Berries

Don't forget about wild berries when you find yourself lost in the wild. Just be careful when looking for berries as some varieties could make you very sick. Stick with berries that you recognize like raspberries, blackberries or other berries that grow in clusters. Other fruits like elderberries can be easily recognized by their purply-black berries that are shaped like an umbrella. Make sure you stay away from any white berries, and all of these will be toxic. Watch out for fruits that look like cherries or blueberries, they may taste great, but they are deadly doppelgangers.

Identifying Poisonous Plants

The best way to identify poisonous plants is by becoming familiar with the kinds that grow in your area like stinging nettle, poison hemlock, poison sumac, poison oak, and poison ivy. Poisonous plants come in many forms, and one rule of thumb isn't going to do it. Poison ivy grows on vines while poison sumac and oak grow in shrubs. Poison hemlock looks like giant bundles of parsley.

The rule I remember as a child is "leaves of three, let it be," but this only applies to poison ivy and oak. Poison sumac will have clusters of between seven and 12 leaves. To make things worse, there are other plants that also have leaves of three like the box elder sapling. This makes it even harder to tell these plants apart. Stinging nettle leaves will be heart-shaped, fine-toothed, and tapers on their ends. Stinging nettles will be extremely hairy; in fact, the whole plant is covered in hairs. They are even on the stems and the underside of the leaves. Even though you can eat this plant, you need to harvest them wearing gloves. If you touch any of the hairs, a chemical gets released that makes you feel like you have a swarm of bees stinging you. Stinging nettle will have very bright pink or yellow flowers. Poison sumac and oak don't have flowers, but poison ivy has clusters of small yellow flowers.

Plants in Your Yard

You have to realize that many of those beautiful plants that adorn your yard could be very toxic to you. Some can cause skin discomfort if you touch them, and you definitely don't want to eat them.

Monkshood is a great example of this kind of plant. It has beautiful stalks of purply-blue flowers that look like little wings. It does look beautiful in your garden, but it is very toxic. Just touching this plant can get the toxins on your skin, which in turn could cause numbness, tingling, and if it seeps into the skin could cause damage to the heart.

You Touched a Poisonous Plant, Now What?

First of all, try to figure out what plant you could have touched; wash the area with cool water and soap. Pat dry the area never rub. Make sure you don't touch any other parts of your body until you have had time to wash the affected area.

If a rash develops that doesn't go away with proper treatment, it can affect your heart rate or breathing, or causes more than just a mild reaction; you need to see a doctor as soon as possible.

CHAPTER 6

List Of Edible Wild Plants

Edible plants can be found anywhere from sidewalk cracks to forests. There is a plethora of free food out there just waiting for you to harvest them as long as you know what you are looking for.

Below you will find some plants that grow wild and are edible:

Alfalfa

This plant is part of the pea family. It can be found in fields across the country. It is very nutritious and has many benefits, like treating drug and alcohol dependency. The young shoots and leaves can be eaten raw.

Asparagus

This plant can be found growing almost anywhere in the world where there are gravelly or sandy conditions. It is dense in nutrients, vitamin C, fiber, calcium, and potassium. It's a great source of B vitamins, phosphorus, iron, magnesium. It can be eaten just like the kind you find in the grocery stores. It can be eaten raw or cooked lightly. You can fry, bake, boil, steam, or sauté it.

Blue Vervain

This plant can be found in most countries but is very abundant in the US and Canada. It likes moist conditions with full or partial sun. The seeds can be eaten when roasted and ground into powder. It can be a bit bitter in flavor. Leaves can be dried and used as a tea or tossed into soups, stews, and salads. The flowers can be used in salads, too.

Broadleaf Plantain

This is a member of the plantain family, and it can be found all over the world. It is full of vitamins K, C, and A. The whole plant is edible, but the young leaves are the best tasting. You can use these in any way that you would spinach-like sandwiches and salads. Some people have been known to eat the shoot of seeds when it has finished flowering. The more mature leaves can be eaten raw, but they are stringy and bitter. If you use the bigger leaves raw, think about taking out the veins first.

Bull Thistle

Bull thistle can be found in various environments but likes areas that don't get disturbed. It grows best where the soil stays slightly moist, but it has been known to grow in both wet and dry soils. You can find it on roadsides, fence lines, waste places, edges of forests, and pastures. Its rosettes can be troublesome in gardens and lawns. You can eat this plant. The young leaves and flower stems can be eating in salads or sautéed. The taste is a bit bland. You HAVE to remove all the prickles from the plant before you eat it.

Cattail

During summer and early fall, you could see lots of fuzzy, brown cattails swaying in the breeze in swampy places all over Canada and the United States. There are many parts of this plant that are edible. Cattails can produce more starch per acre than yams or potatoes. Unlike yams and potatoes, you are able to eat more than just the roots. Various parts of this plant can produce edibles at different stages of their development. The roots can be grilled until tender and eating like an artichoke. If you harvest the brown fuzzy catkins during spring while they are still hidden in the leaves, you can eat them like corn. Boil until hot and serve with pepper, salt, and butter. The stalks and shoots can be harvested and cooked like asparagus. You could clean them and spread on some peanut butter.

Chickory

This is a member of the dandelion family but is a bushy plant with blue flowers. You can eat the leaves and flowers raw, but the roots have to be boiled. Just like dandelion, it can be found all over the world.

Cleavers

Cleavers normally grow around the edges of fields and hedgerows. They could also be found near gardens, woodlands, pastures, disturbed areas, waste areas, orchards, and near crops. This plant is native to western Asia and Europe but is have made its way through North Africa, South America, Central America, Mexico, United States, Canada, and Australia. The stems and leaves can be used as a leaf vegetable even though it is sticky so they won't blend well in salads. You can use them in sandwiches or sautéed. The fruits can be collected, dried, and roasted to be used as a substitute for coffee. The stems and leaves could be dried as a tea.

Coltsfoot

This plant can be found in disturbed, open places. It can be found all over North America. The flowers of this plant can be eaten. You can toss them into salads to give it a great aromatic flavor. You can fill a jar with the flowers and then add honey to make a cough remedy. You can use this infused honey to sweeten any herbal tea. Flowers can be dried and chopped up to put into fritters and pancakes. Young leaves could be added to stews and soups.

Common Sow Thistle

This plant grows pretty much anywhere. It loves being in gravel banks, meadows, fields, roadsides, and cracks in driveways. This plant was brought to North American from

Europe. The roots, flowers, and leaves are all edible. They are best when the plant is young because the older plant becomes bitter. The root can be harvested while young. Roast and grind the roots as a substitute for coffee.

Common Yarrow

This plant can grow in South America, Africa, Australia, Asia, Europe, and North America. The leaves can be eaten cooked or raw. They have a bitter flavor when added to salads. They are best when harvested young. The leaves of this plant can be used in beers. You shouldn't consume these in large quantities. You can make tea from the leaves and flowers.

Creeping Charlie

This plant is native to Europe and southern Asia but was brought to North America to be used medicinally. It quickly adapted to its new surroundings and can now be found everywhere except in deserts and the coldest places in Canada. It is a relative to mint and is hard to control since it easily roots from any node on its stem. It does have a pungent minty flavor and works well when used as a herb. It is best eaten when the leaves are young. You can cook older leaves like spinach. You can also dry the leaves to be used in teas.

Crimson Clover

This plant can be found in most countries and loves growing in cultivated areas, lawns, fields, and meadows. The flowers and seeds are edible. The seeds can be sprouted and then used in sandwiches and salads. You can dry the seeds and grind it into flour. The flower heads can be used fresh or dried and put into teas.

Curly Dock

Curly dock is a great plant to eat as a snack. You can find these in almost every yard or meadow. Even though the leaves can be a good source of vitamins B and A, you need to eat it moderately as it could cause urinary tract infections and kidney stones. The roots of these plants can grow as far as two feet deep, so make sure you don't break them off when digging them up. Once you have them out, you can eat the roots and leaves fresh. It can be dried for later use.

Daisy Fleabane

This plant grows all over North American and has even naturalized to central Europe. The leaves are the only part of this plant that you can eat. They are a bit like hair, so they will have a fuzzy texture when you eat them raw. You can cook them like you would any green. The extract of leaves contains caffeic acid, which has neuroprotective effects on neuronal cells.

Dandelion

Every part of this plant can be eaten. This includes the roots, leaves, and flowers. The dandelion leaves are normally between five and 25 centimeters long and possibly longer. The flowers will be yellow to orange in color. Dandelions can be found all over the world. You can either eat all parts of this plant raw or cooked. They are a good source of vitamins K, C, and A. It also contains folate, B vitamins, and vitamin E.

Downy Yellow Violet

This violet is native in almost every area of eastern Canada and the United States. It also grows in some temperate regions of Asia and Europe. The leaves and flowers are edible. The root used to be used in ancient times for medicinal purposes.

Fern Leaf Yarrow

This plant grows in most countries. Even though the leaves are bitter, they can be consumed cooked or raw. The younger leaves, when mixed in a salad are best. Yarrow leaves can be used in beer making. Even though it is very beneficial and nutritious for any diet, you shouldn't eat too much regularly. You can make a tea from the dried leaves and flowers.

Field Pennycress

This plant can be found all over the world. This plant is edible, but it isn't very tasty. It is from the mustard family and has a very pronounced flavor. The larger the plant, the stronger the flavor, and after they bloom, they can be rather spicy. Cooking will take some of the edges off. You can put the young greens in salads if you want a bit of a kick. If you love eating radishes, you will love this plant.

Forget Me Not

The flowers are the part of this plant that is edible. They can be eaten as a snack or tossed into a salad. You can decorate dessert and garnish your meals with these pretty flowers.

These flowers grow well in zones five through nine in the United States. They can be put into baked goods as well as candied.

Garlic Mustard

This plant can be found anywhere in North American. It is a very invasive plant. The leaves are best when harvested young, and the roots taste a lot like horseradish. The leaves will be a bit bitter as they get older. A second-year plant can be eaten from early spring until about the middle of spring before the shoots have had time to harden. The seeds are great in spicy foods. Using garlic mustard all year long is a great way to keep this plant from spreading. The parts of the plant from older plants do contain cyanide. These parts need to be cooked well before consuming them.

Harebell

This plant grows throughout the US and Canada, along with being in Britain. This plant likes grassy, dry places. The leaves are the edible part of this plant and are best when put into a salad. You could add them to dips, or smoothies.

Japanese Knotweed

This plant can be found across North America. Not every part of this plant is edible. The shoots are what is edible, but they need to be harvested before the stems get woody and hard. The best time to eat knotweed is the middle of April until May. These shoots can be consumed raw and have a sour taste that is similar to rhubarb. You should cook them similarly as you would rhubarb. Think about knotweed in a cocktail, in preserves, or a pie.

Meadowsweet

This plant is normally found by rivers, lakes, wet woods, marshes, swamps, etc. It grows throughout Europe and North American, along with some parts of Asia. The young leaves are edible but are mostly used to make teas and soups.

Milk Thistle

This plant grows in most places around the world. The young flowers, roots, leaves, and stalks are all edible. The roots can be eaten cooked or raw. If you decide to eat the leaves, make sure you remove all the thistles first. The leaves make a great substitute for spinach. The flower bud can be cooked. The stems are best when peeled and soaked to help reduce their bitterness. You can use milk thistle-like rhubarb or asparagus, or you can put it in salads. The seeds of the milk thistle can be roasted and used as a substitute for coffee.

Prickly Pear Cactus

This plant can be found in most desert areas. It is a very popular plant as it is full of carotenoids, antioxidants, and fiber. The edible parts of this plant are its fruit, stems, flowers, and leaves. You can eat the whole cactus, either grilled or boiled. It can also be made into jams and juices.

Purple Dead Nettle

This plant can be found all over the world. This is a very nutritious superfood. The leaves are very edible, with its purple tops being very sweet. Because leaves are a bit fuzzy, so they are better when used as a garnish or mixed with other ingredients in recipes instead of being the star of the show. You can use purple dead nettle along with dandelion greens, chickweeds, or other "weeds" to make a "wild pesto." You can use this green as any other leafy green or herb.

Red Clover

You can find this plant all over the world in meadows, roadsides, grassy areas, vacant lots, pastures, and fields. The leaves can be put into salads or used as a tea. The part that is used the most is the flowers. Red clovers are the tastiest of all the species of clovers even though you shouldn't eat too many as it can cause bloating.

Sunflower

This plant is edible from its roots to its seeds. You can find them all across the United States. You can make everything from teas to salads. The roots can be chopped raw, marinated, mashed with potatoes, steamed, shredded into a slaw, sliced thin and fried, and roasted. The sprouts can be used just like soybean or alfalfa sprouts. The stalks taste a lot like celery. They can be chopped and added to salads or eaten like celery with peanut butter or hummus. The leaves can be used in salads or cooked like spinach. The petals make pretty garnishes, but you can eat them in salads, too. They have a bittersweet flavor that can be used to complement your dishes. Let's not forget about the seeds. The seeds are ready to be harvested when the flowers have turned from green to yellow. These seeds can be eaten raw, or soak them in salted water overnight and roast them for a great snack we all know and love.

Wild Black Cherry

This plant is native to North America. Even though the fruit is edible and can be used when cooking and in beverages, the rest of the plant has amygdalin and could be toxic if consumed. Wild cherry syrup is obtained from the bark and can be used as cough medicine. Wines and jellies can be prepared from the fruit.

Wood Sorrel

This plant looks like a clover but is not in any way related to clover. It is a great thirst-quencher and a great snack. The flowers of the plant can vary from bright yellow to green. The leaves are a wonderful source of vitamin C. All parts of the plants are edible and have a slightly sour taste that can be compared to lemons. This plant can be found all across North America, where there are dirt and sunshine.

CHAPTER 7

List Of Medicinal Wild Plants

Not only can you find a lot of edible plants to cook within the wild, but you can also find quite a few plants that can help heal minor ailments. All of these plants are also edible, but they are also well-known for their healing qualities. While these plants are considered medicinal, always consult with your doctor if you are experiencing any major health problems.

Bee Balm

This is an edible herb and is often used to add flavor to foods, and it makes an attractive garnish for salads. Herbalists use this plant as a stimulant, diuretic, diaphoretic, carminative, and antiseptic. It can be used to help with insomnia, menstrual pain, nausea, flatulence, sore throat, fever, gastric disorders, headaches, and colds.

Birch

The bark off of the birch tree, especially sweet birch bark, can be a great analgesic. The bark can be steeped to make a great tea. If you consume too much of this, it can end up causing nausea, upset stomach, or tinnitus. If you notice any of these things, you should stop taking it. It is recommended only to take 6 to 12 mg of salicin each day. Birch trees are found throughout most of North America and can be spotted by their white bark.

Black Cohosh

This plant is a member of the buttercup family and is native to North America. Black cohosh is used as a dietary supplement for menopausal symptoms and hot flashes along with premenstrual syndrome, cramps, and to help induce labor. It is thought to be used to help treat a range of other illnesses like high blood pressure and tinnitus. The flowers have a fairly strong odor are can repel insects. Large doses of black cohosh can cause poisoning and harm the liver.

Blackberry

Blackberries can be found pretty much anywhere. They are delicious in pies, but they can also help with diarrhea. They can also be made into a mouthwash because of their astringent nature. To make a tea, pour hot but not boiling water over some blackberry leaves and let steep for ten minutes. The normal ration is about two and a half ounces of leaves to one cup of water. You have to drink this when it was made. You can use this same tea as a mouthwash.

Black Walnut

The green hulls on the black walnut are commonly used in folk medicine. A tablespoon of the dried hull can be steeped in a cup of hot water to make a horrible tasting tea. Drinking a cup of this a day for a week can help you rid your body of parasites. The fresh hulls can also be used as a substitute for iodine tinctures on wounds and cuts. Black walnut trees are commonly found in the eastern US.

Burdock

The leaves and roots of this plant make a great liver tonic and can help to purify the blood and body. A lot of people have used burdock root to get rid of acne. It also has positive effects on many skin issues, such as eczema. You can make a tincture of the dried root in some alcohol and take ten to 20 drips every day. The fresh leaves and roots can be eaten after boiling them in water and getting rid of the water to get rid of the bitterness. You can find it around river banks and other disturbed habitats.

Chamomile

Chamomile is a commonly used flower for teas. It is so popular that you can easily find it on the tea aisle at the grocery store. It is a mild sedative and is great for treating insomnia and other nervous conditions. It is safe enough for teething children. It has anti-inflammatory properties and is great for arthritis. It can also help with menstrual cramps. Chamomile can also be used in shampoos and is common in slaves to help heal wounds and hemorrhoids. A compress of it can be used on skin inflammation to help with burns and sunburns. It is an annual herb that originally came from Europe, but can now be found on almost every continent. It likes sunny spaces and can be found from Southern Canada to Northern US and west to Minnesota. There are two types of chamomile, German and Roman. They both have the same healing properties, but Roman chamomile is perennial, whereas German chamomile is an annual.

Chickweed

This edible plant has been used as a folk medicine for hundreds of years for conditions like obesity, skin conditions, dyspepsia, inflammation, constipation, conjunctivitis, blood disorders, and asthma. The extract of this plant can be taken internally. It is normally used externally to treat sores and rashes. You can eat the
young shoots in salads. It can be found throughout North American and Europe.

Comfrey

The cooked and mashed roots of comfrey can be used as a topical treatment for sprains, burns, bruises, and arthritis. DON'T eat it. Research has shown that it can damage the liver if eaten. It contains pyrrolizidine alkaloids.

Echinacea

This is also known as coneflowers. It can be used as an herb and an ornamental plant. It grows best in USDA zones three through eight. Even though all of the plant is edible, the leaves and flowers buds are the parts that are most commonly used to make tea. Echinacea can help reduce inflammation, lowers blood pressure, helps calm anxiety, reduces the risk for breast cancer, help cells grow, controls blood sugar, can help fight the flu.

Elderberry

This tree can be found all over North America. The berries of this plant are very edible. They can be harvested and made into pies, syrup, jams, and wines. The whole flower cluster can be dipped in batter and fried. The petals can be eaten raw or made into a tea. The flowers can add an aromatic lightness and flavor to fritters or pancakes. The berries can be cooked down and made into a syrup. This syrup can help with cold discomforts and as a cough suppressant.

Evening Primrose

This plant can be found across North America. All parts of the plant are edible. It has a mild taste, but some say there is a rough aftertaste. The roots can be eaten cooked or raw, just like potatoes. If you soak the stem in water and boil it, the taste is similar to black salsify root. The leaves can be harvested from April to June before it begins flowering. You can eat them raw in salads or cooked in soup or like spinach. Native Americans make a tea from its leaves as a dietary aid. The oil from the seeds can be used to treat sore throats, digestive problems, hemorrhoids, bruises, acne, eczema, helps with PMS, eases breast pain before periods, reduces hot flashes, reduces high blood pressure, heart health, and reduces nerve pain.

Fireweed

This is a native plant found in the Northern Hemisphere. You can identify it easily by its erect, smooth, red stems. It has unique leaves that have a circular pattern on them that doesn't stop at the end. It has beautiful purple flowers. All those parts of the plant that grows above the ground can be used to make medicine. It can be used for swelling, pain, enlarged prostate, wounds, tumors, and fevers. It could be used as a tonic and astringent. The early shoots can be enjoyed raw or cooked lightly. Harvest the leaves when they are pointing upwards and close to the stem. These leaves can be pinched off and eaten like any leafy green. Once the plants get larger, the leaves get fibrous and unpleasant. The flower buds can be added to salads for a colorful addition.

Ginseng

Ginseng has a long history of being used for medicine that goes back more than 5,000 years. It can help the body adapt to emotional and mental stress, hunger, cold, heat, and fatigue. Ginseng is great for our metabolism, and it can promote well being. It also has a reputation for being an aphrodisiac. These facts seem to be based on the fact that it can relax someone who is very tense. If you have chronic back pain or TMJ, you can add this to a cup of slippery elm and catnip tea.

Henbit

This plant can be found throughout the US and Canada. It can also be found in Greenland, western Asia, South America, and Australia. You can eat this plant cook or raw, along with being used in teas. The leaves, flowers, and stems are edible, and even though this is part of the mint family, most people state that these plants taste a bit like raw kale. It is high in fiber, vitamins, and iron. This medicinal plant can be used as a stimulant, laxative, febrifuge, excitant, diaphoretic, and anti-rheumatic.

Herb Robert

This plant has a history of being a "miracle" plant, but it has been connected to witchcraft. You can eat this plant. The leaves can be eaten raw or steeped in teas. Due to its odor, you can plant a ring of Herb Robert around your garden to keep rabbits and deer away. You can rub the leaves on your skin to repel mosquitoes. It can support the immune system since it is an antioxidant. It can lower blood sugar and can help with diabetes. European herbalists have used it to treat toothaches, bruises, wounds, dysentery, gout, nosebleed, jaundice, conjunctivitis, and cancer. Like all other geraniums, it can help alleviate diarrhea. This herb can also be used to treat foot-and-mouth disease in animals.

Honeysuckle

This has been used as a medicine for thousands of years in Asia. A decoction of honeysuckle stems can be taken internally to treat hepatitis, mumps, and rheumatoid arthritis. The flowers and stems can be used together in an infusion to treat dysentery and URIs. The flowers bud can be turned into an infusion to treat various ailments like enteritis, colds, bacterial dysentery, tumors, and syphilitic skin diseases. Flower extracts can lower blood cholesterol. It can also reduce blood pressure. It can be used as a tuberculostatic, antiviral, and antibacterial. The glowers can be applied straight to sores, infectious rashes, and inflammations.

Jewelweed

If you ever find yourself in contact with poison sumac, ivy, or oak, look for some jewelweed. Crush up the purplish stalk into a paste and rub it over the affected skin. Leave it on for two minutes, and then wash it under clean water. If you can do this within 30 to 45 minutes after exposure is better. You shouldn't have much of a reaction. If it took you a little longer to find some jewelweed, you might still have some relief if you use it as a wash. Jewelweed can also be used to cool the itch on an already established breakout. It can be found in moist, semi-shady areas.

Lamb's Quarters

This plant can be found all around the world. It is full of essential minerals and vitamins, vitamin A, vitamin C, and a good source of B vitamins, including niacin, riboflavin, and thiamine. You can use this plant just like spinach. You can use them fresh in juices, salads, or any recipe that might call for greens. They are best when harvested young, but as the leaves mature, the flavor could change because of the greater potency of oxalic acids. Taste the leaves before you harvest a bunch to make sure their flavor is what you want it to be. You can chew the leaves and put it sunburn, inflammation, injuries, minor scrapes, and insect bites. A tea made from the leaves can help with loss of appetite, stomach aches, internal inflammation, and diarrhea. This tea can be used as a wash to help heal skin problems. The leaves eaten either raw or fresh can help support the blood system and heal anemia.

Lavender

Lavender is normally used as a fragrance these days, but it has been used since ancient times to repel insects. It has also been used to treat skin disorders, burns, and bug bites.

It can help relieve rashes and the itching that goes along with them. It can also reduce swelling. Crush some fresh leaves and apply them to the affected area. You could fill a jar with some dried leaves and cover them with some olive oil. Allow this to sit between six and eight weeks. You can then use the oil for any skin problems. You should never take lavender internally by small children, nursing, or pregnant women.

Lemon Balm

Lemon balm can be found all around the world. You can make the best-tasting lemonade you've ever had by adding some muddled lemon balm leaves to your lemonade. This plant is also good for cold sores and can be used to fight insomnia. Germany's version of the FDA has lemon balm listed as being a better remedy for cold sores than the leading prescription. Crush some fresh leaves and place them over the sores. You can use a cream that contains a very high concentration of lemon balm.

Lobelia

A word of caution, do not take too much lobelia as it can result in death and less-severe side effects. It has been used to help treat respiratory complaints, like asthma, pneumonia, whooping cough, and bronchitis. It is an expectorant, emetic, analgesic, stimulant, sedative, and antispasmodic. It is a perennial herb native to Eastern North America from Maine to South Dakota and south to Missouri and Texas. They bloom from July to November and can grow to three feet high.

Willow

The weeping willow tree is an easy to spot tree common throughout North America. While it is not native, it does well in moist areas. You can spot the tree by its droopy branches and twigs. The bark and leaves have been used to make medicine for centuries. Boiling a palmful of the leaves in a cup of water for ten minutes will make an astringent. Soak a cloth in it to apply to skin problems you are having when you have no other medical treatments available. Bark scraping can be soaked in a cup of hot water for ten minutes and then drunk to stop diarrhea. Take a sip of it every two hours, and continue this until the symptoms go away.

Yarrow

Crushed flowers and leaves can be placed on scratches and cuts to stop their bleeding, and it reduces the chance of getting infected. The leaves encourage clotting, and they are an antiseptic.

CHAPTER 8

List Of Poisonous Plants

Lastly, we will go over some of the most common poisonous plants. It is just as important to know what plants out there will hurt you. Not all plants have the same level of toxicity. In the list below, we will use the numbers 1 through 4 to represent the toxicity levels. One is major toxicity, meaning it can cause serious illness or death. Two is minor toxicity, which can cause minor side effects, like diarrhea or vomiting. Three is oxalates, which means its sap or juice contains oxalate crystals, which can irritate the throat, tongue, and mouth. Four is dermatitis, which means it simply irritates the skin. There won't be very many threes and fours.

African Boxwood

The leaves on this level two plant are the most dangerous. Touching the plant can also cause skin irritation. When eaten, you could experience convulsions, dizziness, diarrhea, vomiting, or nausea.

Angel's Trumpet

A level one plant and the entire plant should be avoided. When ingested, it can cause hallucinations, tachycardia, paralysis, memory loss, and ultimately death.

Azalea

A level one plant and the entire plant should be avoided. While small amounts won't cause much more than a stomach upset, if large amounts are consumed, or if you eat honey made from azalea pollen, it can cause serious problems. This normally happens in areas where there is a dense population of these plants. The honey is often called mad honey because of the confusion it will cause.

Bird-of-Paradise

This is considered a level two plant. The seeds of the flower contain toxic tannins, and the leaves have hydrocyanic acid. It is typically considered non-toxic towards humans and is more harmful to pets, but if large amounts are ingested, it can lead to dehydration or choking.

Bitter Nightshade

This is considered a level one plant, and the berries are the most dangerous. It is a woody perennial. It can reach six feet in height, and its purple flowers will turn into roundish berries. They contain solanine, a poison, which can end up causing headaches, stomachache, drowsiness, lowered temperature, trembling, vomiting, and diarrhea.

Black Henbane

A level one plant, but the seeds and leaves are the most dangerous parts. No part of the plant should be consumed. It can cause salivation, nausea, headache, diarrhea, vomiting, convulsions, rapid pulse, and coma.

Black Locust

A level one plant, and the seeds, leaves, and barks can cause the most problems. Fatal cases of black locust ingestion are rare, but recovery from its side effects tend to take several weeks.

Bushman's Poison

A level one plant and its sap is the most poisonous part. The sap of the plant was often used by the bushmen to create a cocktail that they would dip their arrows in to help them kill when hunting.

Castor Bean

A level one plant and its seeds, leaves, and bark are the most dangerous. This plant is what is made to create castor oil, which is an emetic. While it may not kill you to take castor oil, as long as you take the recommended dose, consuming the plant can, especially the bean.

Checkered Lily

This is a level one plant. The bulb of this plant is the most dangerous. The bulb contains poisonous alkaloids. In fact, all lilies and daffodils contain this dangerous alkaloid.

Chinaberry

This is a level one plant, and its fruit is the most dangerous. Symptoms of ingestion are diarrhea, vomiting, breathing difficulties, and paralysis. Birds and cattle are the only animals that can consume the berries without harm.

Chinese Lantern

This is a level one plant, and its fruit and leaves are the most dangerous. Sometimes referred to as the ground cherry, the pods of the Chinese lantern plant can be highly toxic and fatal.

Climbing Lily

This is a level one plant, and no part of the plant should be eaten. The tubers of the plant resemble yams but are one of the most toxic parts of the plant. It can be fatal.

Coral Bean

This is a level one plant, and no part of the plant should be eaten. Some people say you can eat certain parts of the plant, but it is safer to avoid it altogether. Some parts are hallucinogenic and narcotic.

Daphne

You should avoid the fruit and most parts of this level one plant. Chewing on any part of the berries, bark, foliage, or flowers can prove to be fatal.

Deadly Nightshade

This level one plant is also known as belladonna and is one of the most famous poisonous plants. It contains scopolamine and atropine in is roots, berries, leaves, and stems. The berries start green but turn black when ripe. The tricky part is they are sweet and juicy. A couple of berries can kill a child. An adult would have to eat somewhere between ten to 20 berries before they die, but they are going to be very sick even if it doesn't kill them.

Death Camas

You should avoid all parts of this level one plant. The mature leaves and bulbs are the most toxic. The plant contains several steroidal alkaloids. It can cause ataxia, tremors, muscular weakness, and prostration.

English Laurel

This level one plant's seeds should be avoided, but any part can be poisonous if enough is ingested. It causes potentially fatal respiratory problems.

English Yew

This level one plant should be avoided, but its seeds, leaves, and bark can cause the most problems. It is mainly used as an ornamental plant, but the ingestion of any part can be fatal.

European Mistletoe

No part of this level one plant should be consumed. Ingestion of the plant can result in death. American mistletoe, though it has a common name, is not as poisonous, and will likely only cause minor intestinal upset.

Flowering Tobacco

This is another level one plant. Its leaves are the most dangerous parts. Just like any other tobacco plant, it contains anabasine, nicotine, and other alkaloids. Its toxins are easily absorbed through the lungs and stomach. When too much is ingested, it can cause elevated heart rates, agitation, and coma.

Foxglove

This is a level one plant, and no parts of the plant should be eaten. While it has been used as a heart medicine, it is actually very poisonous. It can be safe in the right doses, but it is very easy to overdo it. Symptoms of poisoning are low blood pressure, collapse, and an irregular heartbeat. You could also experience drowsiness, intestinal upset, rash, lethargy, headache, depression, and blurred vision.

Heliotrope

The seeds of this plant are a level one. The plant can cause gastric distress. It also contains liver toxins, which can end up causing liver damage.

Holly

This is a level two plant, and the berries are the most poisonous parts. If swallowed, it can result in diarrhea, vomiting, drowsiness, and dehydration.

Poison Hemlock

No part of this level one plant should be consumed. This plant is very dangerous for the fact that it looks a lot like other non-poisonous plants. At first, it will cause salivation, nervousness, and tremors, but if it moves into a depressive phase, it can result in death.

Poison Ivy, Oak, and Sumac

These are considered a level 4 because they contain what is known as urushiol. This will cause an allergic reaction that can cause a rash on your skin. It is found every in the US, except for Hawaii and Alaska. They can be spotted by their leaves of three.

Japanese Pieris

No part of this level one plant should be consumed. This plant falls into the same category as azaleas and rhododendrons. If consumed, it can cause abdominal pain, vomiting, and excessive salivation after only six to eight hours after consumption.

Japanese Yew

This is a level one plant, and most parts of the plant should be avoided. This plant contains taxine A and B, which can prove to be fatal when ingested. The main symptoms are vomiting, breathing trouble, and tremors.

Jerusalem Cherry

The fruit of this level one plant is dangerous. This is a nightshade and is known as a pseudocapsicum. Its main poison is salanocapsine, which is a lot like other alkaloids found in this genus. It is not typically life-threatening. It can cause gastrointestinal upset and can cause vomiting.

Jimsonweed

No parts of this level one plant should be consumed. Poisoning most commonly happens when somebody eats or sucks the juice out of the seeds from the plant. It can also be toxic through touch, though not as common.

Lupine

The seeds of this level one plant can be dangerous. This plant is also known as bluebells. It can cause intestinal upset in humans. The poison is found in the foliage of the plant, but it is more concentrated in its seeds.

Monkshood

This is a level one plant and is considered poisonous to the touch. Most of the time, touching the plant won't cause any serious effects; it can cause tingling or numbness in the hand. If eaten, especially the seeds and roots can cause heart problems, tingling sensation of the skin, diarrhea, coma, and possibly death.

Myrtle

This is considered a level two plant. It is also known as common periwinkle. It causes systemic toxicity, which can cause mild abdominal discomfort to serious cardiac problems.

Nephthytis

This is a level three plant. This plant won't necessarily kill you if you eat it, but it can cause a lot of unpleasant symptoms that can occur for two weeks after eating it. It can cause difficulty swallowing, swelling of the throat, drooling, and burning sensation of the tongue, lips, throat, and mouth.

Pokeweed

All parts of this level two plant are toxic to humans and animals. The roots tend to be the most poisonous. The stems and leaves cause intermediate toxicity, and the berries are the least toxic. Children are most commonly poisoned because they will eat the berries. That said, some people will eat pokeweed, but only after the leaves have been properly prepared.

Tansy

This plant is considered a level four. It mainly only causes contact dermatitis for people who are allergic to the plant. If the flowers and leaves are consumed in large amount can cause poisoning because it contains thujone, which can cause brain and liver damage.

Water Hemlock

Not parts of the level one plant should be consumed. This is one of the most poisonous plants in North America. Only a very small amount of the plant can poison humans and livestock. Its main toxin is cicutoxin, which affects the central nervous system.

Woody Nightshade

This level one plant should be completely avoided. This is sometimes called bittersweet. Its egg-shaped red berries and foliage are poisonous. They contain solanine, which will cause convulsions and death if large doses are taken.

Wormwood

This is considered a level four plant. It is part of the Asteraceae family, like marigolds and ragweed, which means if you are allergic to those plants, you may experience an allergic response to wormwood. It is also toxic to the kidneys if too much is consumed.

CHAPTER 9

Recipes For Foraged Plants

Stinging Nettle Spanakopita

You Will Need:

- Melted butter, .5 c
- Phyllo sheets, 18
- Grated nutmeg, .25 tsp
- Chopped parsley, .33 c
- Beaten egg, 2

- Grated parmesan, .5 c
- Crumbled feta, 1.5 c
- Chopped ramps or scallions, .75 c
- Melted butter, 2 tbsp
- Fresh stinging nettle leaves, 8 c

You Will Do:

1. Start by steaming the nettle leaves until they are wilted. Remove them from the steamer and let them completely drain. Place them on a cutting board and chop.
2. Next, add the ramps or scallions with two tablespoons of melted butter in a skillet and sauté. Add in the nettle and sauté for a few more minutes.
3. Set the skillet off of the heat and add in the nutmeg, parsley, egg, parmesan, and feta and mix everything together.
4. Take a 9 by 13 pan and lightly coat it with some melted butter. Unroll the phyllo and cover it with a damp dishtowel. This keeps it from drying out. You will only be working with one sheet of phyllo at a time. Work as quickly as you can.
5. With a pastry brush, brush the butter on the first phyllo sheet. Place it onto the buttered pan, off-centered, so that it covers most of the bottom and comes up one of the sides.
6. Brush your next sheet with butter and lay it on the other side of the pan. Do this six more times, each one of them being placed so that they come up the other side of the pan.

7. Now the next four sheets, once brushed with butter, should be placed in the very center bottom of the pan.

8. Spread the nettle mixture evenly over the phyllo.

9. Now, work as quickly as you can to butter and place the reaming sheets of phyllo on top of your filling. Take the phyllo that is up the sides of the pan and fold them down on top of the phyllo you just places. Brush with butter.

10. Score the top of your phyllo so that it is cut into 12 pieces, but make sure you don't cut down through the filling. Allow this to bake for 35 to 45 minutes at 375, or until everything is golden.

Dandelion Fritters

You Will Need:

- Tallow, lard, or other fat for frying
- Egg
- Melted butter, 1 tbsp
- Milk, .5 c
- Salt, .5 tsp
- Baking powder, .25 tsp
- Flour, .5 c
- Dandelion flowers, 1.5 c

You Will Do:

1. Start by adding your chosen fat to a pot and heat it up so that it is ready for frying.
2. Combine the salt, baking powder, and flower. Mix in the egg, melted butter, and milk.
3. Working one at a time, coat a dandelion flower in the batter.
4. Drop the coated flower into the hot oil and fry until browned, turning them once.
5. Lay them out on some paper towels to soak up the excess oil.
6. These are great with maple syrup or honey.

Seaweed Salad

You Will Need:

- Toasted sesame seeds
- Sliced green onions, 3
- Sea beans, .3 lb
- Various seaweeds, 1 lb

Dressing:

- Sugar, 1 tbsp
- Soy sauce, 1 tbsp
- Sesame oil, 1 tbsp
- Rice wine vinegar, 2 tbsp

You Will Do:

1. Start by whisking all of the dressing ingredients together until the sugar is dissolved.
2. Boil a pot of water and add in the sea beans. Boil for a minute and then place them in a bowl of ice water. Boil the seaweed for 15 seconds and then place it into an ice bath. Dry the vegetables and place them into a bowl.
3. Add the rest of the salad ingredients into the bowl and pour the dressing over top.

Wild Mushroom Ragu

You Will Need:

- Grated hard cheese
- Pepper
- Oregano sprig
- Mushroom soaking liquid, 1 to 3 c
- Dried oregano, 1 tbsp
- Red wine, 2 c
- Tomato paste, 2 tbsp
- Chopped garlic, 4 cloves
- Finely chopped onion
- Diced bacon, .25 lb
- Olive oil, 3 tbsp
- Dried wild mushrooms, 2 to 3 oz

You Will Do:

1. Soak the mushrooms in a bowl of hot water. Cover them so that they stay submerged. They need to soak for at least 30 minutes, or up to a few hours depending on how big they are. Take the mushrooms out once they have been rehydrated and rinse. Place them on some paper towels and squeeze them dry.

2. Pour the soaking water through a sieve so that you get rid of the girt.

3. In a pot, add the butter and bacon. Cook until crisp. Take out the bacon and add in the onions. Cook until they start to brown. Add the mushrooms and garlic, stirring well. Cook for a few minutes until the garlic browns up. Mix the bacon back in along with the tomato paste.

4. Add in the oregano and red wine. Once boiling, let it cook down for five minutes. Add the mushroom liquid.

5. Taste to see if you need some salt. Let this simmer until the liquid has reduced a bit, and the soup is thickened up some. This will take about 20 minutes. Then, turn to low, cover, and cook for an hour.

6. Cook some ragu pasta, and once the mushrooms are done, top the pasta with the sauce. Add some grated cheese, pepper, and oregano.

Stir-Fried Dandelion Greens

You Will Need:

- Pepper and salt
- Olive oil, 1 tbsp
- Red pepper flakes, .25 tsp
- Minced garlic, 1 to 2 cloves
- Fresh, washed dandelion greens, 3 to 4 cups

You Will Do:

1. Add the olive oil to a skillet and heat. Add in the red pepper flakes and garlic, and stir-fry them so that the garlic doesn't brown. Once the garlic has softened, add in the greens.
2. Stir the greens so that they are well coated in the oil. Continue to stir them around for about five to eight minutes. The goal is to wilt them, but not let them become soggy.
3. Transfer to a dish and enjoy.

Purslane Tacos

You Will Need:

- Minced garlic clove
- Diced small jalapeno
- Diced tomato
- Purslane, 1 c
- Small onion
- Coconut oil, 1 tbsp
- Beaten eggs, 2
- Salsa
- Queso fresco
- Warmed corn tortillas
- Pepper
- Salt

You Will Do:

1. Beat the eggs and season them with some pepper and salt. Place to the side.
2. Heat a skillet and add in some oil. Cook the onion for a couple of minutes, or until it turns translucent. Add in the purslane. Stir and cook for two additional minutes.

Add in the garlic, jalapeno, and tomato. Cook everything together for a minute. This will help to cook the juice off of the tomato.

3. With a spatula, push the mixture to the side to make some space to cook your eggs. Scramble your eggs for a few minutes and then mix the eggs and the purslane together. Add some more pepper and salt if needed.

4. Serve the mixture in some corn tortillas and top with salsa and queso fresco.

Zucchini and Purslane Soup

You Will Need:

- Pepper and salt
- Thickener of choice, 2 tbsp – cornstarch or flour works well
- Heavy cream, 1 c
- Chicken broth, 4 c
- Shredded basil, 2 tbsp
- Zucchini, 3 lb
- Washed purslane leaves, 2 c
- Minced garlic, 3 cloves
- Diced medium onion
- Extra virgin olive oil, 2 tbsp

You Will Do:

1. Start by heating the oil in a pot and add in the garlic and onion. Cook for a couple of minutes until they become fragrant. Add in the sliced zucchini and sprinkle in a bit of pepper and salt.
2. Cook until the zucchini has become soft. Add in the purslane and basil. Cook for a few more minutes, stirring often. Make sure that they don't brown.
3. Once cooked, add the vegetables into a blender or food processor and puree.

4. Heat the broth up and add the pureed vegetables into the broth.

5. Combine your chosen thickener with the heavy cream and whisk it into the soup. Cook everything for three minutes, or until the soup has thickened.

6. Taste and adjust any flavors that you need to. Enjoy.

Purslane Salad

You Will Need:

- Pepper and salt
- Zest of a lemon
- Buttermilk dressing, 3 tbsp – use your favorite
- Blue cheese, 4 oz
- Fresh blueberries, 1 c
- Purslane leaves, 4 c

You Will Do:

1. In a salad bowl, add the purslane and drizzle with the buttermilk dressing. Toss to coat the purslane.
2. Fold in the blue cheese and blueberries.
3. Add the pepper, salt, and lemon zest. Stir everything together and enjoy.

Stinging Nettle Soup

You Will Need:

- Heavy cream or buttermilk, 1 c
- Peeled and quartered potatoes, 5 to 6 small
- Chicken stock, 4 c
- Salt, 2 tsp
- Dried thyme, .5 tsp
- Chopped onion, .5 c
- Chopped celery, 1 c
- Olive oil, 2 tbsp
- Nettle leaves, 4 to 6 c

You Will Do:

1. Bring a pot of water to a boil. As you wait, wash the nettles in a sink of cold water, using tongs. Place the nettles into the boiling water, making sure they stay submerged for two minutes. Pour them into a strainer and then run cold water over them to make them stop cooking.
2. Heat the oil in a pot and add in the onion and celery, cooking until the onion is translucent. Mix in the thyme.
3. Pour in the stock and salt. Let this come to a boil. Add in the potatoes.

4. As the potatoes cook, chop up the nettles and mix them into the soup. Simmer the soup for 30 minutes to an hour. Take the pot off of the heat and mix in the cream and buttermilk.

5. Use an immersion blender to make the soup smooth.

6. Add more pepper and salt if needed.

Creamy Nettle Soup

You Will Need:

- Pepper and salt
- Heavy cream, 2 c
- Juice of a lemon
- Zest of a lemon
- Nutmeg, .25 tsp
- Blanched nettle leaves, 4 c
- Chicken stock, 1 quart
- Butter, 2 tbsp
- Russet potatoes, 2
- Sliced and soaked leeks, 2
- Minced garlic, 2 to 3 cloves

You Will Do:

1. Start by mincing the garlic and set it to the side. Slice up the leeks and divide their rings out and soak them in some cold water to get rid of the dirt and said. Drain them. Peel and chop the potatoes.

2. Add some water to a large pot and let it come up to a boil. Add the nettles, working in batches, into the water for 30 seconds. Bring them out and put them straight into ice water. Drain them.

3. In a dry pot, add the garlic, leeks, and butter. Sauté them until fragrant and soft.

4. Add in the potatoes and stock. Simmer the mixture until the potatoes are fork-tender.

5. Add in the lemon zest and juice, nutmeg, and nettle. Bring everything back up to a simmer.

6. Set the pot off of the heat. Using an immersion blender, puree your soup until smooth. You can also use a blender and work in batches.

7. Mix in the pepper, salt, and heavy cream.

8. Serve with some sour cream or grated cheese.

Fiddlehead Soup

You Will Need:

- Pepper, .25 tsp
- Salt, .5 tsp
- Chopped chives, .25 c
- Heavy cream, 1 c
- Diced medium potato
- Bay leaf
- Chicken stock, 4 c
- Minced garlic, 1 clove
- Sliced spring onions, 3
- Butter, 2 tbsp
- Fresh ostrich fern fiddleheads, 1 pound

You Will Do:

1. Start by rinsing and cleaning the fiddleheads well with cold water to get rid of the brown skin. Drain them and set them to the side.
2. Add the butter to a pot and melt.
3. Add in the spring onions and cook them for about three minutes, or until they are tender and fragrant.

4. Add in the garlic and cook for another minute.

5. Pour the stock in and add in the bay leaf. Allow this to come up to a boil.

6. Add the potato and fiddleheads. Turn the heat down to a simmer and let it cook until all of the vegetables are tender. This will take about 15 minutes.

7. Add in the pepper, chives, salt, and heavy cream. Constantly stir the soup until it slightly thickens. Serve with a garnish of green onions.

Dandelion and Violet Lemonade

You Will Need:

- Water, 6 c
- Lemons, 10
- Cane sugar or honey, .5 c
- Fresh violet flowers, 2 c
- Fresh dandelion flowers, 2 c

You Will Do:

1. Begin by adding the water to a pot and bringing it to a boil.
2. Set the water off the heat and add in all of the whole flowers. Let them steep for 20 minutes.
3. Strain your flowers out of the water and add them to your compost, or dispose of them however you want.
4. Let the tea cool down to about lukewarm, and then add the sugar or honey.
5. Pour into a pitcher.
6. Juice the lemons and add this to the tea mixture. This will make the water change to a pretty pink color.
7. Stir together and adjust the sweetness to sit you. If needed, you can add more water to fill up the pitcher.
8. Refrigerate until it is cold before serving.

Ginger, Pineapple, and Purslane Smoothie

You Will Need:

- Diced ginger, .5 tsp
- Turmeric, .5 tsp
- Yogurt, .5 c
- Pineapple, 1 c
- Purslane, 1 c

You Will Do:

1. All you need to do is add everything to a blender and mix until combined. Enjoy.

Purple Dead Nettle Tea

You Will Need:

- Dried purple dead nettle, 1 tsp
- Boiling water, 1 c

You Will Do:

1. Steep the dried purple dead nettle in the water for about 10 minutes.
2. Serve with some honey.

Violet Syrup

You Will Need:

- Violet petals, 1 c
- Water, 1 c
- Sugar, 1 c

You Will Do:

1. Heat the water up to a boil and set it off of the heat. Add in the petals and cover. Let this sit for 24 hours.
2. Using a double boiler, heat up the water and petals and mix in the sugar.
3. Let this come to a boil, stirring often. Set it off of the heat and strain into a clean jar. This will keep in a fridge for six months.

Wild Violet Vinegar

You Will Need:

- White balsamic vinegar, 1 c
- Violet flowers, .5 c

You Will Do:

1. Add the flowers to a mason jar so that they fill the jar about halfway. Pour in the vinegar.
2. Cover with a lid. If the lid is metal, place a piece of parchment between the jar and lid so that the metal doesn't react with the vinegar.
3. Set the jar in a cool, dark place for a couple of weeks.
4. Use just like you would any other vinegar.

Dandelion Root Chai

You Will Need:

- Maple syrup or honey, to taste
- Milk
- Cold water, 3 c
- Whole allspice, 3
- Whole star anise
- Green cardamom pods, 4
- Whole cloves, 5
- Peppercorns, 1 tsp
- Cinnamon stick
- Ginger root, 1 inch
- Roasted dandelion root, 2 tbsp

You Will Do:

1. Start by mixing the dandelion root with the spices in a pot. Pour in the cold water.
2. Allow this to come to a boil and then reduce to a simmer. Let simmer for ten to 15 minutes. Set it off of the heat.
3. Strain the tea.

4. Add the tea to mugs, filling them about ¾ of the way full. Add in the milk to fill up the cup and sweeten to taste.

Fireweed Tea

You Will Need:

- Fresh fireweed leaves, 2 lbs

You Will Do:

1. Strip the good leaves off of the stalks and place them into a bowl.
2. Pick off some of the leaves and roll them between your palms. Set these rolls into a bowl.
3. Once all of the fireweed is rolled, place the lid on the bowl and place them out of sunlight. Let this sit for two to three days. Discard any leaves that mold.
4. Once the leaves are nearly black, place them in the sun to dry. You can also roast them for 20 minutes at 350.
5. Place this in a mason jar. You can now make some tea. Brew just like you would any tea.

Hibiscus Syrup

You Will Need:

- Sugar, .5 to 2 c
- Fresh hibiscus calyx, 20
- Water, 3.5 c

You Will Do:

1. Start by bringing the water to a boil.
2. Take the outer calyx off of the seed pod and place the rest into the pot.
3. After the water starts boiling, turn the heat down and simmer for 20 minutes.
4. Strain the hibiscus out of the water and place the water back in the pot. Add in the sugar and bring it back up to a boil. You can add as much sugar as you want. The more sugar you add, the sweeter it will be.
5. Boil the liquid until it reaches your desired consistency. The longer it boils, the thicker it gets. Five to ten minutes is typically enough.
6. Set the pot off the heat and let it cool. Pour into clean jars and keep it in the fridge for up to six months.

Wild Ramp Pesto

You Will Need:

- Lemon juice, 1 tbsp
- Olive oil, .5 to .75 c
- Sea salt, 2 tsp
- Wild ramp leaves, 6 handfuls

You Will Do:

1. Start by dividing your ramp leaves into the six different handfuls. Two handfuls will stay fresh while the other four will get blanched. This is to cut down on the intensity of the pesto because if you do all six handfuls fresh, you won't have to worry about vampires for the rest of your life. But, by all means, change up the ratio to your taste preference.
2. Add some water and salt to a pot and get it boiling. Working in small batches, blanch the ramp leaves for 20 seconds, or until they become bright green and wilt slightly. Then immediately place them in ice-cold water, and lay them out on a dishtowel to dry.
3. Once you have finished blanching the leaves that you want to blanch, place them along with the fresh ramp leaves into the food processor with the salt, lemon juice, and olive oil. Feel free to add anything else you would like in your pesto right now, such as pepper, walnuts, or parmesan.

4. Process the ingredients until smooth, adding extra oil as needed.

5. Taste and adjust the lemon juice and salt to taste.

Polish Fermented Mushrooms

You Will Need:

- Smashed garlic, 2 cloves
- Cracked pepper, 2 tsp
- Caraway seed, 1 tsp
- Dried dill, 1 tsp
- Crushed juniper berries, 6 to 10
- Pickling salt
- Cleaned mushrooms, 3 to 4 lbs

You Will Do:

1. Boil the mushrooms in some salted water for about five minutes. Drain and allow them to cool on some paper towels.
2. Combine the spices and herbs together in a bowl. Sprinkle a thin layer of pickling salt in the bottom of a non-reactive container. Place a layer of mushrooms on top. Sprinkle with some of the spice mixture. Add some more salt. Continue to layer like this until all of the mushrooms are in the container. Finish with a layer of salt.
3. Lay a plate on top of the mushrooms to weight them down. Place this is a dark, cool place for four days. Check on the mushrooms after the first day to make sure that the mushrooms are submerged in their brine. If they aren't, boil a pint of water with a couple of tablespoons of kosher salt. Once cooled, pour over the mushrooms.

4. After the four days have passed, place the mushrooms and their brine into a clean mason jar. Keep refrigerated. This will keep for a few months.

Garlic Mustard Pesto

You Will Need:

- Salt, .5 tsp
- Extra virgin olive oil, .3 to .5 c
- Grated parmesan, 1 c
- Garlic mustard leaves, 4 to 5 c
- Walnuts, almonds, or pine nuts .25 c

You Will Do:

1. Add the nuts to a food processor and pulse a few times so that they become large crumbs. Add in the parmesan and garlic mustard. Pulse until the leaves are minced, and everything is combined.
2. Continue to pulse the mixture as you slowly pour in the oil. You can eyeball how much you need. The mixture should become shiny and wet.
3. Add in the salt and pulse to mix it in.
4. Use however you would like.

Sorrel Sauce

You Will Need:

- Pepper and salt
- Vermouth or stock, 2 tbsp
- Sorrel leaves, .25 lb
- Unsalted butter, 3 tbsp
- Heavy cream, .66 c

You Will Do:

1. Slice the sorrel into very thin slices.
2. Pour the cream into a pot and let it come to a simmer. This keeps it from curdling once the acidic sorrel hits it.
3. In another pot, add the butter and sorrel. Cook, stirring often until it cooks down and the sorrel turns bright green.
4. Stir in the cream and bring the mixture to a light simmer. It is going to be thick, so you add in the stock or vermouth to help thin it out.
5. Add some pepper and salt, and enjoy.

Fennel Sauerkraut

You Will Need:

- Pickling spices, 3 tbsp
- Crushed juniper berries, 1 tbsp
- Pickling salt, 1.6 oz
- Shredded cabbage, 2.5 lbs
- Sliced fennel bulbs, 2.5 lbs

You Will Do:

1. Mix the cabbage and fennel together.
2. Place a layer of the veggies in the bottom of three-gallon crock about an inch thick. Sprinkle with salt and some pickling spices. Continue this until everything is in the crock. Place a weight on top of the kraut and place it in a dark, cool place.
3. Check on them the next day to make sure there is a brine that is covering everything. If not, boil some water and salt together and pour it into the crock.
4. Let this stay in the cool, dark place for at least a week, but up to a month.
5. Take the weight off and place the kraut in a quart-sized jar. Keep in the fridge.

Country Mustard

You Will Need:

- Salt, 2 tsp
- Water or white wine, .5 c
- Vinegar, 3 tbsp
- Mustard powder, .5 c
- Mustard seeds, 6 tbsp

You Will Do:

1. Grind up the mustard seeds in a coffee grinder or by hand. They don't have to be perfectly ground.
2. Pour the seeds into a bowl with the salt and mustard powder.
3. Pour in the vinegar and water or wine. Stir everything together. Pour into a jar and store in the fridge.

Ancient Roman Mustard

You Will Need:

- Salt, 2 tsp
- Red wine vinegar, .5 c
- Cold water, 1 c
- Chopped pine nuts, .5 c
- Chopped almonds, .5 c
- Mustard seeds, 1 c

You Will Do:

1. Grind up the mustard seeds in a coffee grinder or by hand. They don't have to be perfectly ground. Add the nuts and grind into a paste.
2. Pour the seeds and nuts into a bowl with the salt and water Mix together and let sit for ten minutes.
3. Pour in the vinegar. Stir everything together. Pour into a jar and store in the fridge.

Pickled Blueberries

You Will Need:

- Champagne or white vinegar, 1 c
- Sugar, 3 tbsp
- Salt, 1 tsp
- Blueberries, 1 pint

You Will Do:

1. Place the berries into a pint jar.
2. Boil the vinegar, sugar, and salt together. Pour this over the blueberries, leaving a bit of headspace.
3. Cover and refrigerate. These blueberries will keep for a year like this.

Pickled Fiddleheads

You Will Need:

- Dried thyme, 1 tsp
- Mustard seeds, 2 tsp
- Peppercorns, 2 tsp
- Bay leaves, 2
- Salt, .25 c
- Water, 1 qt
- Fiddleheads, 1 lb

You Will Do:

1. Trim the ends off the fiddlehead.
2. Bring a pot of water to a boil and add plenty of salt. Boil the fiddleheads for two minutes and then place them in ice water.
3. Dissolve a quarter cup of salt into a quart of water. Fill a jar ¾ of the way with some fiddleheads. Cover with the brine. Weigh them down so that the fiddleheads stay submerged.
4. Place in a cool, dark place for two weeks. If you get mold on the top of the brine, that is fine, just skim it off.
5. Divide the spices between the jars of fiddleheads and screw on the lid. Keep in the fridge.

Candied Angelica

You Will Need:

- Sugar, 1 c
- Water, 1 c
- Baking soda, .5 tsp
- Angelica, 1 lb

You Will Do:

1. Cut the angelica stems so that they fit into a jar. Boil a pot of water and add in the baking soda. Prepare a bowl of ice water. Boil the angelica for five minutes and then place it into the ice water.
2. Bring the water and sugar to a boil. Place the stems into jars and pour the hot syrup over them. Let them cool and then screw on the lid. Leave them at room temperature overnight.
3. The following day, pour the syrup into a pot. Let it boil and add in the stems. Boil for a couple of minutes and then everything back into the jar. Cool overnight again. Do this two more times.
4. After the last boil, place the stalks on a rack to cool and dry. Once room temperature, roll them in sugar and keep them in a jar.

Strawberry Dandelion Cake

You Will Need:

- Room temperature egg whites, 6
- Sugar, 1.5 c
- Softened butter, .75 c

Cake:

- Diced strawberries, 2 c
- Milk, .75 c
- Salt, .75 tsp
- Baking powder, 2 tsp
- All-purpose flour, 3 c
- Vanilla, 1 tbsp
- Room temperature egg

Dandelion Syrup:

- Vanilla, 1 tsp
- Honey, .25 c
- Dandelion tea, .5 c

Frosting:

- Milk, 2 tbsp

- Vanilla, 2 tsp
- Berry puree, .5 c
- Sifted powdered sugar, 5 c
- Softened butter, 10 tbsp

You Will Do:

1. Start by getting your oven to 350.
2. Get two 9" cake pans ready by greasing them with some butter and dusting them with some flour.
3. Start by taking the ¾ cup of butter, add the sugar and place them in the bowl of a stand mixer and cream them together. You can also use a hand mixer if you want.
4. Whisk the salt, baking powder, and flour together in a separate bowl.
5. Add the egg whites to a bowl and add in one whole egg. Whisk everything together.
6. Once the sugar and butter have been creamed, carefully add in the egg mixture a bit at a time. Mix thoroughly after each addition.
7. Mix in the vanilla.
8. Mix in a third of the flour mixture and then half of the milk. Add in another third of the flour and then the rest of the milk. Finally, mix in the last of the flour. Make sure everything is fully combined together, scraping down the sides of the bowl if you need to.
9. Stir in the strawberries.
10. Pour the batter into the prepared cake pans, dividing it as evenly as possible. Bake them for 30 to 35 minutes.

11. Once cooked through, cool them on wire racks as you finish making the topping.

12. For the dandelion syrup: Make a cup of dandelion tea by adding dried dandelion flowers into a tea ball and steep it in a half cup of hot water for ten minutes. Take the flowers out and stir in the honey.

13. For the frosting: Start by pureeing your berries in the blender.

14. Next, whip the butter until it is light and creamy.

15. Slowly sift in the powdered sugar until it is all mixed in.

16. Add in the berry puree and the vanilla.

17. You can add just enough milk to get it to the consistency that you like.

18. To put the cake together, place one of the cakes upside down on a plate and brush the top with half of the dandelion syrup. Top the cake with the frosting.

19. Place the second cake right side up onto the first cake and brush it with the rest of the dandelion syrup. Frost and decorate the cake the way you would like. You can even add some edible flowers or sliced strawberries as decorations.

Douglas Fir Poached Pear and Frangipane Tart

You Will Need:

Crust:

- Almond extract, .25 tsp
- Cold water, 3 tbsp
- Cold cubed butter, 10 tbsp
- Salt, .25 tsp
- Sugar, 2 tbsp
- All-purpose flour, 1.25 c

Frangipane Filling:

- Browned butter, 2 tbsp
- Almond extract, .25 tsp
- Slightly beaten eggs, 2
- Sugar, .25 c
- Almond meal, 1.25 c

Douglas Fir Poached Pears

- Cinnamon stick
- Douglas fir needles, .5 c
- Water, 4 c

- Sugar, 2 c
- Peeled whole pears, 2

You Will Do:

1. Start by bringing your oven to 375.
2. Using a food processor, combine the salt, sugar, and flour for the curst. Add in the almond extract, water, and cubed butter. Pulse until the mixture starts to look like wet sand.
3. Place the crust into a tart pan and use the back of a spoon to press the dough firmly into the bottom and up the sides. Let this sit in the freezer for 30 minutes.
4. Place some parchment paper over the crust and add pie weights, rice, or beans onto the paper. This helps the crust keep its shape while baking. Place the pan onto a baking sheet and bake for 20 minutes.
5. Check the crust to see if it looks dry. If it does, then the crust is done. If not, let it cook for another five minutes. Take the crust out and let it rest for five minutes before taking out the pie weight. Allow the crust to finish cooling as you fix the filling for the pie.
6. For the frangipane: Add the butt to a pot and brown it until it just turns golden and is fragrant. Set to the side.
7. Using a food processor, add the almond extract, eggs, sugar, and almond meal. As the processor is running, slowly add in the browned butter until well combined. Set to the side.

8. For the Douglas fir poached pears: in a pot, add in the sugar, cinnamon stick, fir needles, water, and peeled, whole pears. Allow this to come up to a simmer and cook for 20 minutes. You should be able to easily pierce the pear with a fork. Set off of the heat.

9. Take the pears out and place it on a cutting board. Clean off any needles that may be stuck to them. Strain the syrup of the needles and place it back into a pot and let it simmer until the syrup has reduced to two cups. Set to the side.

10. Once you can handle the pears, halve them lengthwise and remove the core and seeds. Trim the stem and blossom end of the pear, and then slice it crosswise into quarter-inch thick slices. Do your best to keep the pear form intact.

11. To assemble the tart, pour the frangipane into the crust, smoothing it out. Carefully pick up the pears, still in the pear shape, and place them onto the filling so that the stem endpoints to the middle. Carefully fan out the slices.

12. Bake this for 40 to 45 minutes, rotating the pan halfway through so that it evenly browns. Once the frangipane is golden, take it out and brush the pears with the syrup you made. Allow the tart to cool completely.

13. Slice and serve with some warmed syrup if you would like.

Wintergreen Ice Cream

You Will Need:

- Chopped semi-sweet chocolate, 4 oz
- Cornstarch, 3 tbsp
- Maple syrup, 3 tbsp
- Wintergreen extract, .5 tsp
- Wintergreen berries, 2 oz
- Sugar, .75 c
- Whole milk, 2 c
- Heavy cream, 2 c

You Will Do:

1. Reserve a quarter cup of the milk for use later. Heat the rest of the milk, wintergreen, sugar, and cream to a steaming point. Set if off the heat, cover, and let it steep for two hours. Once cool, pour into a lidded container and place it in the fridge. You can leave it here up to overnight.
2. Pour back into the pot and slowly heat it back up. Whisk the cornstarch and reserved milk together. Mix into the ice cream base. Mix in the syrup. Stirring frequently, bring the base back to steaming. Then stir constantly for eight to ten minutes.

3. Switch off the heat and cool. Mix in the wintergreen extract. Pour into an ice cream maker and follow its directions for setting up the ice cream.

4. Fold in the chocolate chips. Keep in the freezer until ready to serve.

Mulberry Sorbet

You Will Need:

- Cassis, 2 tbsp
- Mulberries, 5 c
- Water, 1 c
- Sugar, 1 c

You Will Do:

1. Start by removing the green stems off of the berries.
2. Add the water and sugar to a pot and let it come to a boil. Let this simmer for three to four minutes. Set off the heat and let it cool.
3. Add the berries to a blender and pour in the syrup you just made. Blend until smooth.
4. Press this through a sieve to get rid of seeds and stems.
5. Chill in the fridge for a few hours and then pour into an ice cream maker and follow its directions.

Gooseberry Sorbet

You Will Need:

- Sugar, .75 c – you can add more if needed
- Vodka, 3 tbsp
- Gooseberries, 8 c
- Water

You Will Do:

1. Add the berries to a pot and cover with water. Let this come to a boil and cook for two to three minutes.
2. Set off the heat and crush the berries to a pulp with a masher. Don't use any type of blender. Let this sit until room temperature. Pour through a sieve and refrigerate overnight.
3. Strain again, but place a piece of paper towel inside the strainer. You should get clear juice.
4. Take 1 ½ to 2 pints of the juice and sweeten it to taste. Mix in the vodka.
5. Use an ice cream maker and follow its directions to make the sorbet.

Wild Cranberry Sauce

You Will Need:

- Apple pie spice, 1 tsp
- Water, .25 c
- Maple syrup, 1 c
- Cranberries, 4 c
- Zest of an orange

You Will Do:

1. Start by adding everything in a pot and mix it together. Let this come to a boil, and simmer until the cranberries burst and let the liquid reduce slightly. This will take about 20 minutes.
2. Chill to allow it to thicken, and enjoy.

Paw Paw Ice Cream

You Will Need:

- Egg yolks, 5
- Vanilla extract, 1 tsp
- Sugar, 1 c
- Milk, 2 c
- Cream, 2 c
- Mashed pawpaws, 1.5 c

You Will Do:

1. Start by heating the milk, cream, and sugar together until steaming. Mix in the extract.
2. Beat the yolks together. While stirring the eggs, add in a ladle of the cream mixture. Do this one more time, and then pour back into the pot.
3. Stir and heat back up to a steaming point. Once thickened, it should coat your spoon. Turn off the heat and pour it into a bowl.
4. After the custard has cooled off, whisk in the pawpaw until combined. Pour into an ice cream maker and follow its directions. Keep frozen until ready to eat.

Black Walnut Snowball Cookies

You Will Need:

- Powdered sugar
- Stick of butter, cut into cubes
- Pinch of salt
- Grand Marnier, 1 tsp
- Orange flower water, 2 tsp
- Sugar, 2 tbsp
- Chopped black walnuts, 1 c
- All-purpose flour, 1 c

You Will Do:

1. Start by setting your oven to 300. Combine everything together, except for the powdered sugar. Mash everything together with your hands so that it looks like a lumpy meal.
2. Form into balls and place on a baking sheet. Place in the oven for 35 minutes. Let them cool for five minutes. When you can handle them, roll them in the powdered sugar. Once completely cooled, roll in the sugar once more.

CONCLUSION

Thank you for making it through to the end of the book, let's hope it was informative and able to provide you with all of the tools you need to achieve your goals whatever they may be.

The next step is to start gathering the tools that you need to start foraging. Foraging shouldn't be seen as something odd or unnatural. For a long time, foraging was the only way people could find food other than by hunting. If you choose to forage, it will help you save money, and it will give you the ability to consume food that likely has never been touched by pesticides or other chemicals.

It is important that you remember everything we have talked about when it comes to distinguishing plants. Your overall goal should be to make sure that you don't end up consuming a poisonous plant. It wouldn't hurt to make sure you have your phone with you as well as picture references to help you spot plants, especially when you are first starting out. Once you have foraged for a while, you will find it a lot easier to distinguish the good from the bad. Also, make sure you never take more than you need and don't cause severe damage to the environment. You want to make sure that what you take is able to grow back so that you will be able to return in a year or so and take more. You never want to kill a plant.

There is an endless number of ways to use these foraged plants. They make delicious meals, and they can be used to treat various ailments. The possibilities are endless.

Finally, if you found this book useful in any way, a review on Amazon is always appreciated!

Edible Wild Plants

Over 111 Natural Foods and Over 22 Plant-Based Recipes On A Budget

Joseph Erickson

© Copyright 2020 by Joseph Erickson. All right reserved.

The work contained herein has been produced with the intent to provide relevant knowledge and information on the topic on the topic described in the title for entertainment purposes only. While the author has gone to every extent to furnish up to date and true information, no claims can be made as to its accuracy or validity as the author has made no claims to be an expert on this topic. Notwithstanding, the reader is asked to do their own research and consult any subject matter experts they deem necessary to ensure the quality and accuracy of the material presented herein.

This statement is legally binding as deemed by the Committee of Publishers Association and the American Bar Association for the territory of the United States. Other jurisdictions may apply their own legal statutes. Any reproduction, transmission, or copying of this material contained in this work without the express written consent of the copyright holder shall be deemed as a copyright violation as per the current legislation in force on the date of publishing and subsequent time thereafter. All additional works derived from this material may be claimed by the holder of this copyright.

The data, depictions, events, descriptions, and all other information forthwith are considered to be true, fair, and accurate unless the work is expressly described as a work of fiction. Regardless of the nature of this work, the Publisher is exempt from any responsibility of actions taken by the reader in conjunction with this work. The Publisher acknowledges that the reader acts of their own accord and releases the author and Publisher of any responsibility for the observance of tips, advice, counsel, strategies and techniques that may be offered in this volume.

INTRODUCTION

First off, I would like to thank you for choosing this book. I hope that you find it informative and helpful in whatever your goals may be. Throughout this book, we are going to talk about the different aspects of foraging, herbalism, and edible wild plants so that you can enjoy the benefits of eating food in its most natural form.

The world is full of plants. In each area of the world, the plants have adapted to survive their environment, and as such, they have learned how to handle the wildlife of that area. In some cases, this includes humans. While the people in New York may not have the same types of wild plants available to them as people in Texas do, they still have a lot of tasty and healthy options to forage for.

You may be wondering why we are talking about foraging for plants when there are grocery stores on every corner. Well, there may come a time when we can't fully rely on our modern infrastructure to provide users with the necessities we need, especially when it comes to food. But it's not only that. The food you buy in the grocery store has been on the shelves for who knows how long. The produce was picked long before it was fully ripe so that it could travel hundreds of miles without spoiling. While the produce still has important nutrients and minerals in them, they aren't at the same concentration they would be if you picked it off of the tree yourself.

Do you see where I'm going with this?

Moreover, wild plants have medicinal applications and can help with pest control. So not only can you go into your backyard and grab some dandelion greens for dinner, but you also may be able to find some Echinacea to help with your cold. If you got some mint to harvest, you can use that to help keep some pests at bay as well.

You see, nature provides us with an endless supply of possibilities. If you would like to have control over the food you eat, then learning how to forage for food is a great way to do so.

This book will help to guide you through this process. We will go over what herbalism is and what role it played in the history of man. Then we will look at learning more about your environment so that you know what types of foods you can find. We'll also cover the basics of ethical foraging so that you never harm the environment when you go looking for foods. You will also find lists of edible and medicinal plants and how to use them. Of course, you will also find a list of poisonous plants so that you don't end up hurting yourself. Lastly, you'll find some delicious recipes that you can make using your freshly harvested wild plants.

Learning how to forage is a rewarding experience, and I hope you have fun on your first trip out.

Before we get started, I would like to ask that if you find any part of this book helpful or informative, please leave a review.

CHAPTER 1

History Of Herbalism

Using plants for medicinal purposes has been around since ancient Babylonians, Egyptians, Indians, Chinese, and Native Americans. All of these cultures were herbalists. The oldest list of medicinal herbs was called *Shennong Ben Cao Jing* or *Shen Nung Pen Ts'ao,* which is a book about medicinal plants and agriculture. This text is thought to be a compilation of every oral tradition that was written between 200 and 250 CE. It has been said that this text was made from three volumes that contained 365 entries on medicinal plants and a description of each.

The ancient Romans and Greeks were also great herbalists. Surgeons that traveled with the Roman army took their expertise about herbs and spread it through the Roman Empire. They took it into England, France, Germany, and Spain. Galen and Dioscorides, who were surgeons from Greece that traveled with the Roman army, compiled a list of herbs that became the definitive materials for their medical text for about 1500 years.

During the Middle Ages, the monasteries of mainland Europe and Britain preserved herbalism. Before universities were established during the 11th and 12th centuries, monasteries were used as medical schools. Monks would translate and copy the works of Galen, Dioscorides, and Hippocrates. Their gardens were growing the most useful and common herbs that served as training grounds for the following generations of laymen, monks, and physicians.

Because of the Islamic conquest in North Africa during the 7th and 8th centuries, Arabic scholars were able to attain a lot of Roman and Greek texts. The Iranian physician Ibn Sina who was also known as Avicenna, combined the traditions of Galen and Dioscorides with his own ancient practices in a text called *The Canon of Medicine*. It was the most influential text that was ever written. It spread throughout Europe during the 11th and 12th centuries.

Due to the printing press being invented in the middle of the 15th century, the texts of Avicenna, Galen, and Dioscorides were mass-produced and were accessible by people who lived outside of the palace, the university, and the monastery. Using herbs didn't require any special skills. The readers just had to gather their herbs and apply them as described in the text.

Every herbalist who found a new use for an herb tried to standardize the use of this plant. One person who sought to do this was Theophrastus Bombastus von Hohenheim, who was also known as Paracelsus. He emphasized how important it was to get to know a patient rather than just blindly using herbs as a cure.

In spite of his distrust of herbalism, she revived the "doctrine of signatures." According to this, each herb had a particular "sign." Which was how the plant looked, its living environment, scent, or color would show how it was to be used. Any herb that was used to cure jaundice would include dandelions, marigolds, or other plants that had yellow-colored flowers. Pansies that have heart-shaped petals would be used to help heart problems.

One hundred years later, Nicholas Culpeper, an Englishman, brought back another facet of ancient herbalism, and this was astrology. These kinds of herbalists would connect an herb to various signs of the zodiac. They would treat certain ailments by figuring out what planet and sign ruled over a specific body part that needed to be cared for and would then prescribe an herb that had the same astrological sign. Culpeper stated: "he that would know the reason for the operation of the herbs, must look up as high as the stars."

Even though Culpeper and Paracelsus promoted astrological herbalism and the doctrine of signatures, the practice of medicine was changing. Men such as William Harvey and

Francis Bacon were transforming science from being speculative into an experimental process. This didn't mix well with astrology and doctrine of signatures, and so medical and biological science started separating from herbalism. The herbalists who focused on classification wouldn't acknowledge the stars and signatures that eventually formed botany. The doctors who found Harvey's circulation of blood more useful than the planet's movements began what could have been called scientific medicine.

These four herbalists highlighted here were the forerunners of herbal medicine from its beginning to when it changed into medical science.

Arabs Save the Greek Sciences

In the past ten years or so, I have become interested in herbal medicine's history, and I decided to read some of the authors of the day, but I focused on Western European, East Indian, Arab, Greek, and Egyptian. I found a colorful and rich array of cultures, societies, religions, wars, circumstances, and characters that began the medical and herbal traditions that we know today.

The study of herbal, botanical, and medical history was motivated by what I had heard in different herbal and medicine classes and had read in texts about these fields. What upset me the most was that most of the time, these "tales" were spread without even thinking about whether or not they were true. A colleague of mine stated: "herbal legends are rampant in our field."

Without an accurate description of religious beliefs, societal conditions, cultural morays, interactions of cultures, discoveries, events, and so on, we could fall prey to what has been agreed upon, and most of the time misquoted statements about our herbal traditions. Most people just live with these ideas and believe these false conceptions.

If we don't know the roots that herbalism came from, we will just create a myth or story that favors a certain belief about our scenes, views, or the lives we are part of. You could say: "Women have always been suppressed; just look at the way women who were herbalists were killed. This is why we lost our knowledge." Another interesting belief that has been spread around is: "This healing method was given to this tribe and is the oldest form of medicine, and this is the end of it." These statements limit our views, and they normally aren't true and could lead to a rigid philosophic view. Most of the time, these ideas that have been misconstrued are defended with a lot of arrogance, and the people who think like this are not open to discussing anything else.

If we can put these things in a bigger historic picture, it will open us up to be able to look at this healing art in various ways. We won't get trapped in a certain way of thinking. We will continue to be amazed at how determined humans have been and the curiosity that can't be satisfied, along with our ability to forget all the things we have achieved so far.

It's hard to figure out how to start this as this subject is just so large. With all the cultures beginning, ending, and then melding into another culture, wars, religious wars, a culture's forgetfulness, finding the information again, all the superstitions, how the world looks at this, and on and on and on.

Another problem that is encountered when describing this topic is that there are certain areas in history where people have experienced different views. Our experiences today as modern American's are very different from a Syrian refugee's experience. Speaking historically, I have heard people make statements such as: "During the Dark Ages learning about medicine was stagnant." You have to first know what period of time we were talking about, and what part of the world you are talking about. During 600 AD, superstitions ran rampant during Western Europe, but there were a lot of expansions and learning about herbal and medical knowledge within the Byzantine Empire.

I'll begin this historical journey when the Western Roman Empire collapsed during the fifth century. During this period, and for 500 years more, Western Europe had lost a connection with most of its heritage. All that remained of the Greek sciences were Pliny's *Encyclopedia* and Boethius's texts on mathematics and logic. Pliny ended up dying in Pompeii because he was curious about what happened with a volcano erupted. When he made his trip there, he died from the fumes of the eruption. They had such a limited library that theologians found it pretty much impossible to learn more.

Byzantine Empire

In 395 AD, the Roman Empire had been split into West and East because of theological differences. After the fall of the Western Empire, the Eastern Empire decided to claim the Roman world. The boundaries of Rome were shifted, and the Byzantine Empire was centered on the Southern Balkan Peninsula and Asia Minor. During the 1000 years that it existed, the empire was constantly upset by internal political and religious strife and invaders. In spite of this complex administration, moral decay, and gross violence, the empire continued with the Greco-Roman civilization and blended it with the Middle Eastern influences. All this was happening while the West was in complete chaos. During all the years of divisions, wars, turmoil, and being encompassed by the Ottoman Turks, Constantinople fell to Muhammad II in 1453. This began the shit to the modern era. The Byzantine Empire was mainly Green in nature since they spoke Greek, and their main religion was Christian Orthodoxy.

Even though all the turmoil, the medicine of the Byzantine was practiced from 400 AD to 1453 AD. For about 500 years, the herbal and medicinal roots of Western Europe stagnated by superstitions, but within the Eastern Roman Empire, herbal medicine was still going strong. They drew knowledge from Ancient Roman and Greek books that have been preserved in its large library. But medicine continued to be one of the few sciences that the Byzantines were better at than their Greco-Roman predecessors. Because of this, we are able to see the way their medicine influenced Arabic medicine, as well as Western medicine's rebirth during the Renaissance. Anytime the word medicine is used, it will be referring to all kinds of herbs.

The physicians in Byzantine would compile all their knowledge into books. They would elaborately decorate these books with illustrations showing a certain ailment. Paul of Aegina wrote *The Medical Compendium in Seven Books*. This book was written during the late 7th century AD. It stayed in used as a textbook for over 800 years.

There was a revolution in medicine, and there are several sources that mention hospitals being established. Constantinople was probably the center of all of this during the Middle Ages due to all of their knowledge, their geographical location, and their wealth.

Ancient Greek Medicine

This will be a very brief overview. There isn't time or space to tell you everything about this era of medicine.

Byzantium and Western Europe medicine were created by the Ancient Greeks medicine, and this was influenced by Egyptian and Babylonian traditions. Hippocrates, a Greek physician, created humoral medicine that was made up of four humors, phlegm, black bile, yellow bile, and blood. He used the elements in the nature of air, fire, water, and earth to get these back into balance.

Hippocrates has been called the "father of modern medicine." There are around 70 medical works that were created from Hippocrates and his students. This collection is known as The Hippocratic Corpus. He also came up with the Hippocratic Oath that physicians still take today.

Hippocrates and his followers were the first to start describing many different medical conditions and diseases. He had gotten credit for his critical thinking within the medical world, and for finding the cause of disease through logic and observation. He did his best to get rid of superstitions and metaphysical causes of diseases. He found these ideas to not be effective, and they didn't help relieve the conditions of his patients.

The next authority on plant-based medicine was Pandonios Dioscorides in both Western Europe and Eastern Arab Empire. Dioscorides tried to come up with a good field guide that contained plants that can be helpful in medicine. He wanted to discover a way to retrieve that kind of information that he needed in order to treat people. Dioscorides was one of the first people to say: "Anyone wanting experience in these matters must encounter the plants as shoots newly emerged from the earth, plants in their prime, and plants in their decline. For someone who has come across the shoot alone cannot know the mature plant, nor if he has seen only the opened plants can he recognize the young shoot as well." He gained information from traditions and knowledge and mentioned others like Theophrastus. Theophrastus had tried to learn about all plant families. He believed plant superstitions should not be used and tried to place plants into categories that had some sort of order. This information was split between five books:

- Sharp aromatic herbs
- Pot herbs
- Cereals
- Shrubs and trees that produce raw materials that can be used medically
- Resinous and oily plants that can be used to make aromatic ointments and salves

He also talks about how important juices and roots of the plants are, too. He has listed the seeds that can be used medicinally and created an inventory of all the cordials and wines that he used to treat his patients.

Later in Rome, Galen, another Greek physician, was a great surgeon during the ancient world and performed various surgeries, which included eye and brain operations that nobody tried again for nearly 2000 years. He wrote more about herbs and was able to identify herbs that grow in various locations. He also created another version of the humoral system of medicine that was more rigid but did contribute to more inquiry about surgery, wound healing, and herbal medicine.

Early Middle Age European Medicine

During the 5th and 9th centuries in Europe, all their medical knowledge came from the surviving Roman and Greek texts that had been preserved mostly in monasteries. All the ideas about the cure and origin of diseases weren't completely religious but had been based on a spiritual view where factors like the will of the gods, curses, possessions, demons, astral influences, sin, and destiny took more precedence than the physical cause. One author called this "social consensus" and "shamanistic complex." The effectiveness of these cures was based on their belief in the doctor and patient instead of based on empirical evidence. The new view became centered in the Church that forced their influences on medicine.

The center of the new view became the Church that forced their influences on medicine. Since Christianity emphasized compassion and cared for the sick, monks ran the hospitals, but they didn't function like hospitals do today. These were areas where people who were seriously sick were taken. These people were either expected to get better or die according to God's will. There weren't many physicians to tend to these people, just kind monks who gave these people comfort, various herbal medicines, and the sacraments of healing.

Since the Church looked at caring for the soul as more important than caring for our bodies, physical cleanliness and medical treatments were not valued. Being able to deny all bodily desires was viewed as being saintly. With time, almost all Western Europeans looked at illnesses as any condition that was created through a supernatural factor that could end up becoming a diabolical possession. Because of this, cures were only able to be given through religious means. During this time, the Church had a lot of influence on every aspect of people's lives.

Every malady would have a patron saint that prayers were said to that was sent up by the community, friends, family, and patient. Upper respiratory infections could be prevented through blessing the person's throat with candles that had been crossed on the feast of Saint Blaise. Saint Roch was the patron saint of plague victims. Saint Nicaise helped to protect against smallpox. Kings were views as people who were divinely appointed, and many thought that had the ability to cure skin diseases, scrofula, and other maladies by using the "royal touch."

In early Greece, China, India, and Egypt, this type of thinking was widespread, and rituals and incantation were done to ward off evil. Pandemics, bad weather, crop failures, other natural disasters were thought to be caused by either one person or a group of people who had displeased the gods. These people had to perform some sort of atonement. These beliefs are still prevalent in some communities.

In today's times, these ceremonies and prayers could be soothing and could help with the healing process but might not alleviate the condition or illness. Most of the time, a diagnosis from a physician was needed for problems that won't go away and were serious. Without knowing the cause for the condition because of prayer only, a wrong diagnosis, or not getting a diagnosis, the illness could get worse or be prolonged. This isn't trying to debunk using positive thoughts, prayers, or intentions, but relying only on these could do a huge disservice to the patient.

During this time, licensed medicine completely disappeared. The doctors who were still around were usually connected with an abbey or monastery. Even for these people, their goal wasn't to find the cause of the illness or even to heal it, but to study the work of other people. The Catholic Church banned surgery done by monks during the middle of the 7th century since it could damage their souls. Because nearly all surgeons at the time were monks, this decree put an end to all surgeries in Europe. Critical thinking wasn't taught or understood during this time. Aristotle, Hippocrates, and others had made a lot of progress in terms of critical thinking, but all of that ended up getting lost to the Western Europeans. All the talks about life and illnesses were based on theology with little to no scholarly influence. This was a very dark time for Europe.

Arab Influences

Let's move on and see how Arabs were able to save and expand on the sciences and medicine of the Greeks. During the time that Western Europe was falling into a black hole, the East experienced a surge in intellectual life.

Let's start this story in the beautiful and ancient city of Jundi Shapur. During 490 AD, their benevolent king had given shelter to the excommunicated Nestorian Christian scholars that had come from Europe. Many of these people were physicians. These people were Nestorius' followers. He was a patriarch that was tough during the Roman Empire. All these people were excommunicated due to heresy.

Before they settled in Jundi Shapur, they lived with the erudite monks in Syria. This area was meant to be a Macedonian colony. Edessa was located along the north edge of the Syrian plateau. It was known for being in the center of the "Silk Road to China." There was a medical school in Edessa where Nestorians worked and taught.

Zeno, the emperor during 489 AD, said it was a place that was full of heretics. He had the hospital closed, and they went on to Nisibis. They relocated to Jundi Shapur, which was located north of the Black Sea. The people who lived here welcomed the new arrivals after the closing of Plato's school in 529. Some of the Nestorians moved into China and India.

Since the Nestorians were scholars, they started the huge task of translating Greek books into Syriac. Galen and Hippocrates were part of their translations. These adventures helped to explain the reasons some of the Greek work was translated into a Latin version from Syriac. This is why Albucasis, Avicenna, and Rhazes, the great Arabic physicians, revered the Greek teachers and added their knowledge to East Indian medicine.

Many people haven't ever heard of Jundi Shapur, but this was where a lot of knowledge was promoted and preserved. Once the city feel in 636 to the Arabs, and Persia finally became part of Islam, they didn't disturb the university. The people who conquered it adopted it and used it as part of their training center.

Jundi Shapur's rulers welcomed all the Hindu, Jewish, Persian, and Greek scholars. All these people converging created a huge center of medical knowledge within the Islamic world for several hundred years. All of these people of various creeds worked together in harmony. This hasn't happened anywhere else in the world.

These Arab physicians made sure that they understood the work of Galen, Hippocrates, and all of the other prominent Greek physicians, and they were exposed to knowledge of the Chinese and Indians. Intellectuals, astronomers, physicians, and scholars of all branches of knowledge were asked to write, study, and debate about their works in this setting.

Near the end of the 10th century, Baghdad, which had become the capital of the Islamic State, started to drain the talents of Jundi Shapur. This end came extremely fast. Now, nothing remains of that wonderful city except for some trenches in the ground.

It needs to be said that most of these people used plants and exchanged plants from various localities and countries. This helped to expand these materials to other countries. The Islamic Empire soon stretched all the way from North Africa, Central Asia, Italy, Spain, and the entire Middle East. This allowed them to exchange plants.

Baghdad

Knowing how important it was to translate the Greek works into the Arabic language to make it more accessible, Harun al-Rashid and son, al-Ma'mum, created a bureau in Baghdad. This was named the House of Wisdom. This brought a better era to Arabic medicine. We can still feel the effects of this today. This was thought of as the great period of compilation and translation. This boom in development, inquire, and experimentation about all types of sciences was different from what Western European countries were doing. This was the Golden Age for the Arabic people.

The best translator during this time was Hunayn ibn Ishaq al-'Ibadi. He was said to have received a payment of gold for the works he created. He, along with his team, had translated the works of Dioscorides, Hippocrates, Paul of Aegin, Oribasius, and Galen, among others, into Arabic by the end of the 800s. Their writings created the foundation for unique Arab medicine.

Arab medicine followed the theory of humors created by Galen and believed the body was created from the same elements that make up the world: water, fire, air, and earth. All of the elements could be mixed in different ways could be used to help with the different humors and temperaments. If a body's humors are balanced correctly, the person will be healthy. Any sickness was caused by an imbalance in humors and not from a supernatural force. These imbalances could easily be corrected by a doctor's healing art.

Arabic physicians started looking at medicine as science where the body's temperaments could be recognized, and their goal was to preserve the health of the human body. If a person's health was lost, they had a way to help recover it. They looked at themselves as practitioners of the maintenance and healing of the body.

Advances had been made in many other areas. Harun al-Rashid created the first hospital, well, hospitals in the way that we know them now. In about a couple of decades, there were 34 more hospitals built throughout the Arab world. This number increased every year. These were amazing hospitals.

These hospitals didn't resemble the hospitals in Europe. The sick viewed hospitals as a place where they received treatments and could be cured by the doctors there. The doctors viewed the hospital as a place that was devoted to promoting health, curing disease, and expanding and distributing medical knowledge. Medical libraries and schools were soon attached to the larger hospitals, and the top physicians would teach the incoming students. These students were expected to use what they had learned to help the people in the hospital. These hospitals treated everyone no matter what their economic status or religion was. They had readings of the Koran, music, and gardens, and instead of giving the patients a large bill for their services, they provided the poor with money when they left the hospital to ensure that they had a better life as they were healing.

Hospitals also tested the students and would give them diplomas. By the 11th century, the hospital staffed traveling clinics. This provided medical care to people who didn't live close enough or were too sick to get to the hospital. The hospitals were the center of Arab medicine and were the prototype that modern hospitals were based on.

Just like hospitals, herbal pharmacies were developed by the Arabs. Islam teaches: "God has provided a remedy for every illness," and all people are able to search for all of these remedies and use them wisely with compassion and skill.

Jabir ibn Hayyan wrote one of the first pharmacological texts. He is considered the father of Arabic pharmacy. This text was very extensive and gave plants geographical origin, physical properties, and ways to used the plant to cure diseases. Arab pharmacists introduced many new herbal drugs to hospitals in the known world. Some of these new herbal drugs were mercury, ambergris, aconite, cloves, nutmeg, tamarind, cassia, myrrh, musk, sandalwood, camphor, and senna.

These pharmacists also created juleps and syrups. The word julep comes from the Persian and Arabic languages that mean sweet water that is made from things like orange blossom and rose water. They knew all the anesthetic effects of the Indian herbs that could be inhaled or added to liquids.

During this time, the pharmacist was a profession that was only done by people who have been trained extremely well. They had to pass several exams, be licensed, and had to be monitored by their state. During the start of the 9th century, Baghdad saw its first herbal apothecary shop. Pharmaceuticals were created and distributed through commercial means, and then given by pharmacists and physicians in various types like inhalants, suppositories, tinctures, confections, elixirs, pills, and ointments. Herbal medicines were beginning to reach a new level of usage and were being appreciated during this time in history.

Quick Review

In the beginning, the Arabs studied, compiled, and translated the ancient texts from Greek to Syriac, and to Arabic. They created learning centers where people from Western Europe, China, North Africa, India, and Persia would come together to exchange ideas about science and medicine.

By 800 AD, the boom of medical information had been infused with the original thoughts from Arabic medicine. This fusion could really be seen when Al-Razi started focusing on medicine. Some of his most well-known books are an encyclopedia of 25 books called *The Comprehensive Work* that would be translated into Latin. His entire life was spent finding information for his book. The book was supposed to be a summary of medical knowledge. He added his own observations and experience. He told physicians to pay more attention to what their patients and medical history could tell them instead of them just consulting the past authorities. His clinical skills were matched by him being able to understand human nature, especially as seen in a patient's attitude.

He said patients and doctors have to establish mutual trust. He thought that getting positive comments from a doctor could encourage patents, make them feel better, and speed up their recovery. But he did warn then that changing doctors could hurt a patient's health, time, and wealth. This principle about the patient-doctor relationship needs to be looked into by our current medical professionals. Herbalists need to remember the way their clients have built their trust up with time as we get to know their world.

Central Asia Avicenna

It wasn't long after Al-Razi's died that Abu 'Ali al-Husayn ibn 'Abd Allah ibn Sina, a Persian, was born in Bukhara. Later on, translators would Latinized his name to Avicenna. He soon became the Aristotle or da Vinci to the Arabs. His interests embraced statecraft, poetry, music, psychology, mathematics, science, astronomy, philosophy, and medicine. The physicians of his time called him "the prince of physicians."

His father was a tax collector, and he was so brilliant that he had memorized the Koran by the time he was ten. He then studied philosophy, physics, mathematics, and law. At the age of 16, he started to focus on medicine, which he believed was not that difficult. By the time he turned 18, he was famous as a physician, and he was asked to help heal the Samanid prince. His treatment was successful. Due to this, he was allowed access to the royal Saminid library. This was the best library of learning.

He created a huge body of texts that are known as The Arab Golden Age, where all the translations of every text from the Aristotelian, Mid- and Neo-Platonic, and Greco-Roman eras were talked about, revised, and developed by Arab intellectuals. During his life, he wrote about 450 texts on many subjects, and about 250 of these have survived. About 40 of his texts concentrate on herbs and medicine, and 150 concentrates on philosophy.

He is regarded as a pioneer in aromatherapy due to the invention of extracting essential oils and steam distillation that he used during his practice.

His biggest contribution ended up being the encyclopedia named *The Canon of Medicine,* which originated about 1025 in Persia.

This text is known for introducing the idea of syndromes when diagnosing certain diseases, clinical trials, introducing experimental medicines, introducing quarantines to limit spreading diseases, finding sexually transmitted and contagious diseases, studying physiology, and systemic experimentation. Avicenna was one of the first to think about microorganisms and described and classified many diseases. He also outlined their causes. Functions of various parts of the body, complex medicines, simple medicines, and hygiene were covered in his book, too. He was the one that figured out the tuberculosis was contagious. The Europeans tried to dispute this, but it was found to be true. He also discovered the complications and symptoms of diabetes.

Please remember that whenever you see the word medicine in this chapter that all these medicines were minerals, herbal plants, and parts of animals that were used to create medicine. Don't confuse this word with modern medicine that is mostly chemical compounds. His book included a definition for over 760 medicinal plants and all the various medicines you can get from that plant. During this time, he laid out the basics for clinical drug tests, and his principles are still being followed today. He was a wonderful herbalist.

His book became a standard reference for the Arab world. Up until the 1800s, his book was used as a teaching guide and reference. It was used much longer than most other medical books. The healing system he used was called the Unani Tibb that translates into "Medicine of the Greeks." This is something that still gets practice in the Middle East, Central Asia, and India.

The Arab scholars of the 12th and 13th centuries continued to elaborate and develop medicine farther than where the Greeks left off. The traveled far, drew from their own observations, and took expeditions just to find and identify plants. In their texts, we can find coral, elecampane, lesser celandine, euphorbia, horsetail, elder, umbellifera, Artemisia, chamomile, and teasel. It would be several centuries before this type of work would take place in Western Europe.

Reawakening of Europe

In the 1100s, Europe started to receive benefits from the Arabs work, and through this, tried to find its own heritage. Roman and Greek works had been preserved and then expanded upon by the Arabs and were finally making its way back to Europe. Hippocrates and Galen's work came back to the West by the North Africans and Middle Easterners. All the Arab medical texts had been translated into Latin, which was the language of the educated Europeans. Due to the Arabic intellects and scholars, Europe was finally recovering part of its past.

There were two big-time translators who worked to translates the texts from Arabic to Latin. One was Gerald of Cremona; he worked in Toledo. The other was Constantine the African. He was a very learned man and spoke three languages fluently. He was located in Salerno and worked in the cloister of Monte Cassino. It wasn't an accident that these men both lived within this transition zone. This is the area when the two cultures started to influence one another. It isn't a coincidence that Salerno was close to Sicily in Arab. Salerno became the first medical school during the Middle Ages.

Avicenna's text appeared in Europe at the end of the 1100s, and it had a dramatic impact. After being coped and recopied, it was soon set as the standard medical reference in Europe. It had 16 editions, and after several more years, it would pick up more than 20 editions. From the 1100s through the 1600s, it was considered the pharmacopeia of Europe. In 1537, The Canon became the required text for the University of Vienna. This is where herbalism in Western Europe found its roots.

Europeans thought of Al-Razi and Avicenna as some of the best authorities on the teachings that dealt with medicine, and the portraits of these men can still be found in the School of Medicine at the University of Paris.

The Arabs didn't just provide a line of medical knowledge within the Hellenic and Greek world. They also made sure the work was correct and expanded upon it before they passed it onto Europe, who had abandoned the old ways of observing and experimenting that had been created years before. Physicians of various religions and languages created a structure that can still be seen in the medical and herbal practices of today.

Herbalism continued on from this point, and, in a way, we are all herbalists and indebted to the people of these various times, regions, religions, and cultures from all over the world. Just saying that "herbal knowledge was stunted during the Middle Ages," or any other simple statements are being shortsighted. This has a lack of scope, diminishes its brevity, and is completely incorrect. We need to be more educated in our herbal history to see that in one part of the world where it might become stagnant, but there are other areas that are in their glory.

What exactly is "herbalism?" Haven't we been practicing herbalism for years without knowing it? How many herbs became known centuries ago in lands far away from us? Our herbal roots didn't die out with the Dark Ages, now where were all the herbalists tortured or killed. Our roots travel much deeper and into the richer critical thinking, curiosity, and intelligence of the Arabs, Persians, and Byzantine scholars. From all of these people, herbalism was created.

Where did we get elecampane, myrrh, nigella, ginger, thyme, rosemary, oregano, fennel, nutmeg, clove, and cinnamon come from? What could we herbalists do if we didn't have all of these medicinal plants that have come from all over the world?

Because humans are very forgetful, we might have created some limited philosophies about herbs, such as the things that we may or may not use and how we view ourselves in a historical context as herbalists, or how we envision ourselves as part of this multi-cultural heritage could end up clarifying our role in history. This can end up helping up to be more open to fresh ideas of what exactly being an herbalist means. Once we are able to embrace our history and not make up an imaginary history that has been based on rumors, we can finally see ourselves in a bigger scope for all human endeavors to know our place in the world with plants and each other.

CHAPTER 2

Knowing Your Environment

The first thing you need to have a good understanding of is your environment, especially your climate. This will help you to have a good understanding of what types of plants are native to your area. This way, you don't end up spending your entire day looking for a certain plant only to find out it doesn't grow where you live. We will focus mainly on North America.

North America can be divided into eight different climate types:

- Forest – has four different seasons with cold winters and warm summers.

- Coniferous forest – dry and cold with snowy winters and some warm summers.

- Mediterranean – warm temperatures with a lot of rain in the winter and fall.

- Grassland – cold winters with rain and hot summers.

- Tundra – winters are very cold, summers are warm with some rain.

- Alpine – snowy, cold, and windy. It stays winter from October to May, with temperatures staying below freezing. Summer lasts June to September, with the temperatures reaching around 60 degrees Fahrenheit.

- Rainforest – high temperatures with lots of rain all year.

- Desert – warm temperatures with very little rain.

Different plants grow naturally in these various climate types. Let's take a look at where these climates exist within North America.

Forest Land

Forest, which is sometimes referred to as deciduous forest, is located in the Eastern, United States. This stretches all the way up to the tip of Maine and down through Florida, except for the very tip of Florida, which is part of the coniferous forest. It also reaches across to the Eastern edge of Texas and up to Minnesota. All of New England and the South Eastern US is deciduous forest land. Parts of the Great Lakes area are also deciduous forest land.

The deciduous forest also includes part of Southern Canada. These lands are made up of broad-leafed tree forests. These trees shed the leaves each year, which is why this area is referred to as a deciduous forest. But this area also has some conifer trees, which include evergreens.

This area once occupied around 2,560,000 km2. It was dominated by hickories, chestnuts, and oaks that provided food and shelter for the wildlife in the area. During the 18th and 19th centuries, there were unprecedented changes that took place here. The forest was cleared for urban expansion, fuelwood, timber, and agriculture.

The deciduous forest if a type of "temperate deciduous forests." These types of forests occur across the world in areas that are mid-latitudes, the area that occurs between the polar and tropic regions, in Southwestern South America, Eastern Asia, and Western Europe, and Eastern US. They can easily be distinguished by cold and warm air masses that create four seasons each year. Trees will change colors and will lose their leaves in the fall as the temperature levels drip. Winters will often have low precipitation levels and colder temperatures, and plants and trees will become inactive. During the spring, precipitation and temperature levels will start to rise, which causes the plants to come out of their dormancy and new flowering and growth to start. During the summer, plants in this area will grow the most because they are fueled by the warmest temperatures and most precipitation during the year.

Coniferous Forest

The coniferous forest covers most of Canada, as well as Alaska and the Northern part of Washington state. Coniferous forests are mainly found throughout the temperate climate of the Northern Hemisphere. Around 85% of all coniferous forests are located in North American and Eurasia. It extends across the northern part of North America, through Russia and Scandinavia, and across Asia from Siberia to Mongolia, northern China, and northern Japan.

These forests are mainly made up of spruces, northern pines, Douglas firs, silver firs, larches, and hemlocks. These softwood forests create a resource of great importance and yield the bulk of pulpwood and lumber handled commercially.

Mediterranean

The Mediterranean climate is most commonly associated with the Mediterranean Basin, but there are other regions that have a Mediterranean climate. This includes the coastal area of the Western United States, mainly the outer edge of California, as well as the Western Cape of South Africa, central Chile, coastal areas of South Australia, and southern Western Australia.

This climate is characterized by dry, hot summers and wet, cool winters. These regions are normally located around 30 and 45 degrees latitude north and south of the equator. By and large, they are always located on the Western sides of the continents.

The annual temperature ranges in these areas are typically smaller than those in marine west coast climates because areas on the western sides of continents aren't positioned well to receive the coldest polar air, which is developed over land and not the ocean. They are drier than subtropical climates, with their precipitation totals normally range from 14 to 35 inches.

Grasslands

Grasslands are mainly found in the middle of large landmasses. Most of the mid-western part of the US is grasslands, along with the eastern part of Mexico. Grasslands are also found in the steppe that straddles Asia and Europe. Most of this clime is located between 40 and 60 degrees north or south of the equator.

This climate has a large temperature range between the hot summers and cold winters since this region is pretty far from the moderating effects of a sea breeze. Large temperature variations can occur in one place within a single day. Temperatures have the ability to change as much as 30 degrees from day to night. This difference is only beaten by the hot deserts. It has a rainfall total that ranges from 10 inches to 20 inches each year, which makes it a much wetter than the desert.

In North America, these regions are known as the Prairies. In Argentina, they are called the Pampas. In Asia and Europe, they refer to this region as Steppe. In South Africa, it is part of the Veld, and in New Zealand, it includes the Canterbury Plains.

This area has the best soil in the world, making it an important part of food production. This is why this land is normally covered in farmland. The one thing you won't see a lot of in this area are trees because it is too dry.

Tundra

The tundra is located in regions just below the ice caps in the Arctic and extends across North America, Asia, Siberia, and Europe. Around half of Canada and a lot of Alaska are part of the tundra. The tundra can also be found at the tops of the tallest mountains in the world. This area is characterized by extreme cold, but it can get warmer during the summer.

The winter in the tundra is cold, dark, and long, and temperatures stay below freezing six to ten months out of the year. The temperature gets so cold that there is pretty much a permanently frozen ground just below the surface that is referred to as the permafrost. This is the defining characteristic of the tundra. During the summer, the top layer of soil will thaw just a few inches down, which gives vegetation a chance to take root.

The amount of rain that tundra sees is less than the greatest deserts in the world, coming in at around five to ten inches. Even still, the tundra tends to be very wet because the low temperatures cause evaporation to be slow.

Alpine

The alpine climate is made up of mountainous areas. The Rocky Mountains, Alps, and the Andes are all alpine climates. The alpine can be found at an altitude of 10,000 feet, and where the snow line of the mountain begins. During the summer, the average temperature of this area ranges from 40 to 60 degrees Fahrenheit. During the winter, the temperatures typically stay below freezing.

As the altitude increases, temperatures typically fall. The temperature in this area is dynamic and can transition from warm to freezing in a single day. The winter is typically from October to May, and the summer is from June to September. This is a fairly dry climate, with the average rainfall being around 12 inches.

Rainforest

Rainforests are normally located across the world at the equator and are mostly found between the Tropic of Capricorn and Tropic of Cancer. This band is 3000 miles and is known as the tropics. You can find rainforests in Central and South America, Oceania, Africa, and Asia. Rainforests make up around seven percent of the surface of the Earth.

The largest rainforests are found in South America in the Amazon River Basin, in West Africa in the Congo River Basin, and throughout most of southeast Asia. You can also find smaller rainforests in Australia, Madagascar, Central America, India, and other tropical locations. Tropical rainforests have two seasons, the wet season and the dry season.

There are also temperate rainforests that you can find on the Pacific coast of the US and Canada, as well as Norway, Scotland, Ireland, Chile, Tasmania, and New Zealand.

Desert

Lastly, there are desert lands. One-fifth of the Earth is made up of desert land. A desert is characterized by a layer of soil that can be stony, gravelly, or sandy, depending on the type of dessert. They typically get about 20 inches of rain each year, and the things that live there have adapted to this dry climate.

Plants in the desert have learned how to conserve water. Cacti have enlarged stems so that they can store water, and their spines help protect them from thirsty animals. There are four types of deserts. They are hot and dry, cold deserts, coastal deserts, and semi-arid deserts. In hot and dry deserts, which are often called arid deserts, the temperatures tend to be warm and dry all year.

The most famous arid deserts are the Sahara Desert that makes up most of the African continent, and the Mojave Desert found in the Southwest part of the US. Semi-arid deserts tend to be a little cooler. The dry, long summers are typically followed by winters that get a little bit of rain. These deserts can be found in Asia, Europe, Greenland, and North America.

The Coastal desert is a bit more humid than other deserts. While they do get heavy fog that blows in from the coast, it is still rare for them to get any rainfall. The Atacama Desert of Chile is a popular coastal desert.

Cold deserts are dry, but their overall temperature is lower than most other deserts. The Antarctic is the most popular cold desert.

Now that we have a good understanding of the different climates that can be found in North America, let's take a look at what plants you can look for in each.

Edible Plants In Northeast US

The northeastern part of the United States is made up of nine states. They include Pennsylvania, New Hampshire, Connecticut, Rhode Island, Massachusetts, Vermont, New Jersey, New York, and Maine. If you refer back to the different climates that we just discussed, you'll notice that all of these states are a part of the deciduous forest climate.

You will notice that there is some overlap in plants in each of the regions of the US that we will discuss, as well as the climates. The one thing that you have to take into consideration is what area in the state you live in if you live in this area. While their climate is deciduous forest, cities, like New York City, have been so overtaken by people, you'll have to travel out of the city to really be able to forage for anything.

We know that the deciduous forest is characterized by all of its deciduous trees. These trees also provide homes for other plants here. Climbing vines, especially poison ivy, use the trunks as a support system. Mosses and lichens grew on the outer bark of the trees.

Below all of these trees are flowers and shrubs. The forest is broken into several layers of growth. The first is known as the shrub layer, where shrubs and bushes like rhododendron, holly, and azaleas grow. Shrubs tend to be deciduous and will lose their leaves during the colder months. Below this layer is the herb layer where wildflowers like

Dutchman's breeches, trillium, and bluebells grow. They normally grow in early spring before the trees have all of their leaves. Lastly, mosses, fungi, and lichens grew on the ground layer and take in all of the nutrients of the wet soil.

With that in mind, let's look at some edible plants that are easy to find in the northeast region of the United States. Most of what we will look at here are easy to spot, harvest, and consume. Also, most of them are beginner-friendly and hard to confuse for poisonous plants.

Cattail Root

Pretty much anybody will be able to recognize cattails. These plants grow plentifully in marshy areas, so they are most commonly found next to water sources. Its tall stature and distinctive tufted tops make them easy to see from a long way off.

You can pull the plants up by the roots and harvest the tubers. They can grow quite large, and some can even weight around a pound. These tubers can be dried and ground into flour, boiled, or roasted. They have a very mild flavor, so if it doesn't have a mild flavor, then I'd suggest you stop eating eat.

Dandelion

Everybody knows how to spot dandelions. Chance are, you have spent the better part of your life trying to kill them in your yard. They may even magically pop up through your sidewalk. If you don't know what they look like, go outside and look for the yellow flowers in your yard that will eventually turn into puffballs.

The flowers and leaves on the plant are the tastiest parts and can be cooked or eaten raw. Make sure you remove the green stem of the flower because it doesn't taste good. While dandelions can be bitter, especially when harvested in late spring and during the summer, many foragers are used to eating whatever they can find and think they are delicious.

Wild Garlic

Wild garlic is easier to spot during the early spring. It is one of the first plants that grow each year, so it isn't hidden behind other grasses. If you go only by how it looks, wild garlic can be confused with other species, even some that are poisonous. How to really know if it is wild garlic is to go by its smell. If you crush part of it and sniff, you are going to know you have picked the right plant. Both the leaves and bulbs can be eaten.

Watercress

That fancy stuff you buy to put in your salad can be found in the woods. Watercress is a peppery tasting green and is thought to be a delicacy. Chefs desire this green so much they pay people to hunt for it. You can find watercress in freshwater streams. You can eat the stems and leaves. Keep an eye out for floating plants that's leaves are comprised of small oval-shaped leaflets. When you harvest them, cut the leaves off and allow the roots to stay intact so that the plant can grow again.

Ramps

Ramps are coveted as well. These are wild onions, and they have a small but very tasty bulb and leaves. While they are hard to find, ramps grow in patches, so you will have a lot at your disposal when you find them. They are very sensitive to overharvesting, so only take what you need. You could even just harvest the leaves.

Nettles

The one thing that most people know about nettles is that they sting. This probably makes it seem strange that you would want to forage them. However, the sting goes away after they are cooked. Nettles are fairly easy to find, cook, and taste great. They do sting, so make sure you have some heavy-duty gloves when you harvest them.

Raspberries and Blackberries

A lot of people will tell you not to eat unfamiliar berries in the wild. You should listen to those people. Again, don't eat berries you are unfamiliar with. That said, raspberries and blackberries are familiar. You have likely already learned what they look like and how they taste from trips to the store. If you are confident in recognizing them, these tasty berries can be the highlight of your foraging expedition. Once you have spotted the fruit, you can enjoy the leaves on the blackberry plant.

Japanese Knotweed

You will have to forage for this in April. It grows into thick stalks that look a lot like bamboo and tastes like a cross between lemon and celery. Japanese knotweed is a rare opportunity in the foraging world. When you find it, even late in the season, it is going to regrow. This regrowth will give you another chance to harvest the plant in seven to ten days. But, Japanese knotweed is an invasive species that can damage local ecosystems. So eat all you want because you are helping the environment.

Goldenrod

This is fairly easy to spot because of its bright yellow color and the fact that it is several feet tall. It contains long, lanced leaves, and you can find it on the edges of most fields. It is most common during late summer and early autumn. The leaves and flowers are edible and can be eaten raw. You can also brew it in a tea. Make sure you turn the leaves over because the leaves are prone to toxic mold growth.

Milkweed

These plants grow in fields and in large quantities. When it is broken, it will ooze out a sticky substance that resembles milk. It has two distinctive features. It is the monarch butterflies' favorite meal, and it grows large, spiny, edible pods. These pods can be stuffed or boiled, and have an asparagus type flavor. Make sure that you pick the pods when immature, or you could get a mouthful of seeds and fluff.

Edible Plants in Southwest US

The southwest is mainly made up of Arizona, Nevada, New Mexico, and Utah. It also includes the Navajo Nation. This area is mainly desert and is home to some of the US's most amazing natural wonders, including Carlsbad Caverns, Arches National Park, and the Grand Canyon. This region is home to a mixture of different influences, including Anglo, Latin, Hispanic, and Native American traditions.

For those who have never lived in a desert climate, you might be quick to believe that you won't be able to find anything to forage there. If you ever find yourself stranded in the desert, you may just be able to survive if you can learn some of the most common edible wild plants. You'll be surprised that they are nutritious and relatively plentiful.

Tepary Beans

Tepary beans have something that is rare in foraging, a whole lot of protein. It is also one of the most drought-tolerant legumes, with edibles, flowers, beans, and leaves. You can find it growing up to three feet tall on the ground or climbing up something else. Keep an eye out for broad-leafed plants with long pods that look a lot like green beans. Don't harvest from the plant if it is immature since you will only want to eat the beans inside of the pods and not the pods themselves.

Pinon Nuts

You've hit the jackpot if your find pinon trees in the fall. Pine nuts are delicious and full of precious fats. To harvest them, lay a tarp out under a tree and shake the trunks to release the seeds from the cone. This may take a bit of work, but it will be worth it. You can harvest them in late summer, as well, but they will require you to dry or char the cones in order to get rid of the resin that seals the seeds in during this time of year. Pinon trees only produce a decent harvest every few years, so it's possible that you may spot a tree only to find that it doesn't have any goodies.

Yucca

This edible wild plant grows all throughout the southwest, but make sure you don't confuse it with yucca, which is an edible plant in Latin America. The yucca is a spiky, low growing shrub. It can not only be used as food, but it can be used as cordage and soap. You can eat its young stalks, seeds, fruits, and flowers.

Saguaro Fruits

You can easily spot saguaro cacti as they look like the cacti that all of the cartoons depict. The early summer is the best time to get saguaro fruit, but you can find fruit at any time of the year. Fruits are typically at the top of the cactus. You should only eat the interior. They have a refreshing taste. You will want a stick or something to knock the fruit off of the cactus.

Prickly Pears

This is another cactus with edible fruits. It isn't tall, so it won't be as difficult to harvest, but it is obnoxiously spiny, so you got to watch out for that. Not only that, but they are covered with glochids, which are tiny and hard-to-see spines that are nearly impossible to get off of your skin.

Make sure you never touch any are of the plant with your bare hands until the spines have been removed. You can do this by cutting out the nodes or burning them.

Purslane

Purslane is a succulent that does very well in sunny places, which is pretty much all of the Southwest. You can eat every part of the plant that grows above the ground. Keep an eye out for squat shrubs with round leaves and a reddish stem. It tastes like watercress and is better when tender and young.

One other plant that can be found in the southwest is one that we have already talked about, and that is nettle. Nettle can be found everywhere in the US except for Hawaii.

Edible Plants in Southeast US

The southeastern area of the United States consists of West Virginia, Virginia, Maryland, Tennessee, South Carolina, North Carolina, Mississippi, Kentucky, Georgia, Florida, and Alabama. Again, this area is made up of deciduous forests, except for the southern-most tip of Florida. That means all of the plants that we talked about in the Northeast section can also be found in the Southeast. But let's look at a few more.

Persimmon

This is also sometimes called the date plum. It is sweet when ripe, but tends to be sour otherwise. If it is green, it is not ripe, so you should leave it on the tree for a bit longer. The persimmon is orange-red to yellow-orange fruit. They tend to be on the small side, about a half of inch to four inches in diameter. Persimmons can be found in pinelands, dry woods, fields, meadows, and clearings.

Arrowhead

As the name suggests, this plant has leaves shaped like arrowheads. Attached to this is the tuber of the plant and looks like a potato. Since it looks like a potato, you can treat it like one as well. It is peeled, diced, and roasted. If you can't peel it, you can eat it with its skin on as well. No part of the plant is toxic. Arrowheads can be found in shallow water and canals. One of the best ways to find it is to wade in the water, and once you find the plant, just pull on it.

Wild Strawberry

We all know what strawberries look at. You can easily spot the flower before the strawberry is produced. The flower has white five petals and is about ¾ of an inch wide. The petals are attached to a yellow cone. This cone is what will start getting larger and thicker, before turning red when it is ready to be picked. They often grow in woodland borders, slopes, fields, and meadows.

Edible Plants in Northwest US

The northwestern part of the US is made up of Wyoming, Montana, Idaho, Washington, and Oregon. These states are a mixture of different climates, including coniferous forests, grassland, alpine, Mediterranean, and desert. That said, there are quite a few plants that you can harvest in this region. In fact, all of the plants we have talked about so far can be found in this area, even the plants from the desert. Let's take a look at some more plants that you can find in this area.

Asparagus

This plant will need to be harvested in the early spring when it is in its spear form. After a few days, it will continue to grow into what looks like a pine tree. The shoots taste best when cooked. Make sure you don't harvest from a plant that is younger than two years old so that the plant can get established. It is most commonly found growing in disturbed fields and near the road. Young plants can sometimes cause contact dermatitis.

Burdock

The entire burdock plant can be eaten, including the roots. You can eat the young leaves raw. The older leaves taste better when they have been boiled in water. The roots of the first year plant can be added to stir-fries or soups. You can also use the roots just like you would use a potato.

Catnip

While you may have only thought of this plant as something your cat goes crazy over, it is also a great plant for humans, but without the psychotic effects. It is part of the mint family and can be eaten raw or cooked. It is great as a seasoning or as a tea. It grows in disturbed dry areas.

Chickweed

This plant is common in most areas and can be spotted by a single line of fine hair that runs between the stem node. It tends to grow in lawns and disturbed spaces in montane and low regions. The plant is better when it has been cooked and tastes like spinach.

Chicory

This plant is better when you harvest them young, or it is growing in areas that are protected from direct sunlight. If the plant is older, cook the leaves in several changes of water. You can also eat the flower heads on young plants. The roots can also be eaten. Ground chicory roots are often used as a coffee substitute. Look for chicory in disturbed ground, from the foothills and the plains to montane areas.

Edible Plants in Midwest US

The Midwest region is made up of Wisconsin, South Dakota, Ohio, North Dakota, Nebraska, Missouri, Michigan, Kansas, Iowa, Indiana, and Illinois. The climate of this area is mainly grassland, but there are areas that fall into the deciduous forest climate.

This area is well known for its farming because of its fertile soil. It is common in these states to see fields of oat, wheat, or corn growing. That means there won't be a shortage of crops in this area of the United States when it comes to foraging.

Serviceberry

This is a perfectly balanced sweet fruit. These are best picked between June and August. There are several different types of this fruit, and they can be found on bushes, or, occasionally, on trees. None of them are poisonous, but some of them tend to be a bit dry and tasteless. When they are red, they resemble crabapples, and the bottoms of the fruit look like a dried-up flower. The leaves are smooth, serrated, oval, and alternate on the stem. The flowers will bloom in spring, then closer to summer, you will start to see the fruit. They taste kind of like a blueberry and a cherry.

Yarrow

This is a perennial herb that can be found in many different colors, from purple and pink to white and yellow. It grows to be two to three inches in height, with leaves that look very thin and hairy. They are shaped like a lance and cut into small segments. The leaves grow in a basal rosette and are alternate on the stem. It will bloom during the summer, starting in June throughout July. It likes sunlight, so it is most commonly found in open fields.

Wild Bergamot

This is a perennial herb that is part of the mint family. It has a two-lipped tubular flower on square stems. They often grow two to four feet tall and will bloom from July to September. They range in colors from pink, red, or lilac. Wild bergamot will often grow in woodland edges, clearings, and dry thickets. It prefers dry to slightly moist conditions.

Edible Plants in South Central US

The south-central part of the US is made up of Louisiana, Arkansas, Oklahoma, and Texas. This area is mainly grassland with some deciduous forest and desert. This area has a mixture of the plants that we have also discussed along with a few others.

Wood Sorrel

This plant has small, heart-shaped leaves and stems that have a sour taste. It can easily be added into salads or soups for an interesting flavor. You should make sure you don't eat this plant in large quantities.

Turks cap

The edible flowers on the mallow plant are sweet, kind of like honeysuckle, and the leaves taste best when they have been cooked because they aren't as tough. A dark red fruit will grow after the flower has fallen off and is known as the Mexican apple. It has a texture and taste similar to an apple. Turks cap has to be cultivated for gardens.

Autumn Sage

The orange and red variety of these plants all have edible, fragrant leaves and flowers that can be eaten cooked or raw. They are often steeped for tea or used to season food. It grows wild and is very commonly cultivated for landscaping.

Pink Evening Primrose

The stems and flowers of this pretty pink plant can be cooked like greens or tossed into a salad. They taste the best when they are harvest before the flower has a chance to bloom out.

This is not an exhaustive list of the plants you can find in each region of the US. As you have probably figured out, a lot of these plants will grow in just about any area, especially nettle, purslane, and dandelion. We will discuss many more plants in the coming chapters so that you can have a better understanding of what you can look for on your next foraging adventure.

CHAPTER 3

Compendium Of Edible Plants

We've already discussed quite a few different plants that can be found in different areas in the US. But to make sure you are prepared for your foraging trips, this chapter will be a list of edible plants that you can hunt for on your next outing.

Agave

Agave can be found in areas where the soil is sandy or loamy and well-drained. They do not like shade and tolerates drought. The flowers, sap, basal rosettes, and stalks can be eaten. The buds and flowers should be boiled or steamed before consumed. The flowers are great battered and fried. The leaves can be diced and baked or roasted. They have a rich caramel flavor.

Alligator Weed

Alligator weed can be found in California, as well as southern and southeastern states. They grow near water sources and form thick mats. They are invasive, so it is a rather abundant weed. It can be treated like spinach, but don't eat it raw. It has to be cook in order to kill off any aquatic parasites. It is full of minerals and a decent amount of protein.

Amaranth

Amaranth, also known as pigweed, can be found in most areas of the US and likes disturbed areas like edges of woods, yards, and fields. It tolerates most soil types, but it prefers rich soil. It grows near lambs quarters. Pick the smaller to medium-sized leaves because they contain more nutrients. Amaranth can be cooked like spinach, steamed, or sautéed. You should never eat his raw.

Apple

Apple trees can be found in any state that is within the deciduous forest climate. When it comes to foraging for apples, you can either pick them off of trees or pick of fallen apples. A tip, always cook any fallen apples you pick to make sure it is safe to consume. There are many ways to cook with apples. You can turn them into cider, apple butter, pies, and even your own apple cider vinegar.

Arrowhead

Arrowhead is a perennial aquatic plant. It grows fairly well in any area with a decent amount of water, such as streams, marshes, and ponds. The tubers of the plant can be eaten once boiled or roasted. They are typically roasted, and while the skin is edible, they are better when peeled.

Asparagus

Asparagus can be found in every state in the US, but it is still a rare plant to find. They have to be harvested in early spring. Depending on how quickly a state warms up, you can find them as early as February or as late as June. It does not like excessively wet soil but will grow close enough to water to get its benefits. It can grow to six feet in height and is ferny and resembles dill or fennel. You have to be quick at cutting the spears because they can send up early spears that fern out before you even find them. This asparagus is just like the kind you would buy in the store, so cook them however you would typically. They are great roasted and diced up and added to a frittata.

Bamboo

Bamboo is not native to the US, but has been introduced in many different areas and can become invasive. They can grow to 100 feet tall. Look for it in moist, warm areas, mainly within the southeastern US. Some are cold-hardy and can live in the north. Cut the bamboo close to the soil, or dig around the younger shoots, and cut them off just above the rhizome. Clean and then peel the outer sheaths. It is best when harvested in spring. Bamboo can be sliced into coin shapes, boiled, and cooked into stir-fries, soups, or meats.

Basswood

Basswood is full of fiber. The leaves can be used just like greens. The young shoots are also tasty. They can be eaten raw or cooked and can easily be used in place of lettuce in a salad. The flowers can be used to make tea, but they also produce a nectar that can create a great honey. Basswood is mainly found in deciduous forest areas.

Bee Balm

Also known as wild bergamot, it like sunny, dry areas as it can develop mildew easily in overly moist areas. It is a drought-resistant plant. The leaves and flowers can be eaten and are commonly used to help treat colds. It can be cooked or left raw. It is great in salads or as a tea.

Beech

The beech tree produces beech nuts, which was, at one point, a popular food source, but people have hardly heard of them now. They are found in the fall and are full of healthy calories. They have a spiky exterior husk that will pop open once they are ripe, which reveals two small nuts. The seeds will have a fibrous inner shell that can easily be removed once dried and cured. Beech trees can be found in the eastern US. The nuts can be used just like you would use any other nut.

Beautyberry

Beautyberries can be found all along the south and the southeastern US. They grow on shrubs that are about three to five feet tall. They are understory plants and are located in most wooded areas, especially when the soil is moist. Once they are fully ripe, the beautyberry can be eaten. They should be dark purple but not wrinkled. They have a slightly medicinal flavor. They are best when you use them to make a jelly. They can also be used to make wine. Limit how much you eat, though, they have been known to cause mild intestinal distress.

Birch

Different varieties of the birch tree can be found in the western and eastern United States. As an ethical forager, you should only take twigs, leaves, and catkins so that you don't kill the tree. They are minty and peppery in taste and rich in vitamin C. Birch is commonly used to make teas.

Bittercress

Bittercress grows in basal rosettes, and each stalk has around five to nine leaflet pairs. It is hardy and frost-tolerant and will stay green throughout most winters. Tiny white flowers show up during spring and will continue to bloom until fall. Bittercress should be used as quickly after harvest as possible because it wilts. It can be used in salads, soups, on sandwiches, and incorporated into hot dishes.

Black Walnut

The black walnut tree can be found in nearly every state in the eastern part of the United States. The walnuts are easily spotted by their large green shells. It's easier to grab the nuts when they fall to the ground, but make sure you get them before the squirrels do. You can use a knife to cut open the green hull, or you can place them under the wheel of your tire and run over them. Either way, be careful because the hull lets off a sap that can stain. Once out of the green hull, you can let them dry, or you can break open the shell now and remove the nuts. Use the nuts however you like to use walnuts.

Blackberry

Blackberries can be found all over the world. There are several different species of this plant, and they are all edible. There are a few species that don't have thorns, and you would be lucky if you came across these. When foraging for blackberries, be expected to carry home some scratches with you as well. It seems the largest berries are always right in the middle of the thorniest part of the plant. The leaves of the blackberry plant can help with digestive problems such as diarrhea. When the leaves have been dried, it makes a great tea. Ancient Greeks used it to treat gout and problems with the throat and mouth. Blackberries are full of vitamins and antioxidants but will mold if not kept in the refrigerator. Don't wash them until ready to use, as this will promote mold. These tasty berries begin ripening as early as June, but most of the time, somewhere near the end of July. If you pick unripe berries, they will not ripen once taken off the vines. Blackberry seeds are a great source of calcium, Omega 3 and 6, iron, phosphorus, and potassium.

Bladderwrack

Bladderwrack can be found on shorelines in central and northern California and along the northern coasts of North Carolina. It is a perennial seaweed that has a hard, flat root. The frond can range from a couple of inches in length to four feet. It is about an inch wide and has a flat rib that runs down its length. The "air sacks" will vary in size from the size of a pea to a marble. It is best to collect during low tide. It can be eaten cooked or raw. You can also dry it for later use. It has a very strong salty fish flavor. Bladderwrack contains fiber, protein, polyphenols, oleic acid, iodide, phosphorus, iron, manganese, selenium, magnesium, zinc, beta-carotene, B-vitamins, vitamin C, essential fatty acids, silicon, mannitol, cellulose, chlorophyll, zeaxanthin, lutein, potassium, sodium, bromine, mucilage, algin, and iodine.

Blueberry

Foraging for blueberries can be a gateway for foraging for other foods. These blue beauties are easily recognized. They are normally smaller than what you find in stores. They are scrumptious and seedless. These bushes won't prick and scratch you to death like raspberries and blackberries. They are full of manganese, fiber, vitamin C, and antioxidants. Wild blueberries are better than the cultivated ones. Wild blueberries are smaller and have a more intense flavor. There are two kinds of wild blueberries: low brush and high brush. Both of these can be found near low wet woodlands and lakes. They can be found anywhere in the world. Different countries call them different names.

Burdock

The roots of the burdock plant are very medicinal. Most people don't realize that you can eat burdock, too. The roots, stalks, and leaves are all edible and very tasty if you know how to cook it. Burdocks can be found at the edges of most walking paths where the seeds can grab onto animals and humans. Burdock is a biennial, so the roots can be harvested during its first fall or second spring. It is best to harvest the plant while they are young as a mature plant's taproot will be several feet long and about two inches in diameter. They can be quite difficult to dig up. Burdock can be found all over the world, and most people say it tastes just like eating dirt. It is normally peeled, thinly sliced, and added into stir-fries. Some people in the eastern part of the world just wash it, adds it to a pan of olive oil with some salt and pepper, and roasts it. They don't peel it. Burdock tincture is a great anti-inflammatory. Burdock vinegar is great for digestive issues. Just put some roots into a jar and cover with apple cider vinegar. Let it sit for a few months and strain. The "pickled" roots can be eaten.

Cattail

This is a very versatile plant that grows wild in most parts of the United States. This is a water-loving plant that grows on the edges of swamps, marshes, streams, and lakes. If you harvest this plant at the right time, the bottoms of the stems can be eaten raw or cooked. You need to make sure that the water isn't polluted. Never harvest cattails if there is a lot of human activities like riding horses, etc. There could be parasites or harmful bacteria in the water. These are best when harvested in early spring, but if you want to harvest the pollen, it is best to do this in May and June. Other than using this plant for food, the Native Americans use it to make woven baskets, sandals, roofs, and hats. The dried leaves can be used to make children's toys and dolls. The roots can be crushed and put on cuts, bruises, or burns the help with pain and to speed up healing. All parts of this plant can be eaten.

Cherry

Wild black cherry trees can be found throughout the eastern part of North America and into the lower parts of Arizona and New Mexico. They can be found south in Mexico and beyond. They have become naturalized in Europe. The black cherry tree can be used for landscaping, woodworking, and food. The inner bark can be turned into cough syrup. This fruit is important to the survival of many mammals and birds. Many songbirds eat black cherries during migration. Even though the fruit is edible, the seeds, bark, twigs, and leaves are poisonous to humans and livestock. They have a cyanogenic glycoside that will break down when digested that creates hydrocyanic acid or cyanide. Livestock gets poisoned when they eat any leaves that have wilted. Black cherries contain 17 antioxidants. It is a good source of melatonin and contains cadmium, zinc, selenium, copper, sodium, potassium, phosphorus, iron, magnesium, calcium, vitamin C, B vitamins, and vitamin A. It is still debatable if the seeds are edible or not.

Chestnut

The American chestnut is native to North America, but most of these plants were killed out by a fungus that was introduced during the late 1800s from the import of Chinese chestnut trees. About three billion chestnut trees were killed during this time. This tree still thrives in Appalachia, New Jersey, and Michigan. The nuts from both the American and Chinese chestnut are edible. The American chestnut will have their nuts encased in a spiny bur. These will open up and drip their nuts right before the first frost. They will contain two skins, a leathery outer hull that is brown and a papery like skin under this one. This was a very important food source for the early settlers, and most animals like turkeys and deer rely on chestnuts to get them through the winter.

Chickweed

Chickweed isn't a native plant to North America. The European settlers brought it over. Most gardeners hate it as it can quickly take over a garden, but it does contain more minerals and vitamins than kale or spinach. Chickweed likes rich, soft soil that is damp and cool. All parts of the chickweed are edible, the flower, buds, leaves, and stems. You need to be a bit selective when harvesting since only the top two inches of the stem can be eaten. Lower than that, and the stem is too fibrous and stringy. Just some scissors are all you need to harvest this plant. Chickweed is tender and can be eaten right from the plant or added to salads. It is a good source of potassium, magnesium, calcium, and vitamin C.

Chicory

Chicory is best known for its roots, but its flowers and leaves can be eaten, too. It can be found all over the world. It is normally harvested during the fall. Their roots can sometimes be hard to harvest since they like growing in packed soil. If you want to collect the roots, try to find areas where the soil is looser, but this might not be possible. Digging around the roots will increase your chances of getting a lot more. The root is edible and a great source of prebiotics and fiber, as well as insulin. It is normally roasted and used as a substitute for coffee. The leaves can be used just like dandelions in salads or cooked like spinach. The flowers can boost your immune system and could help you relieve some stress.

Chinkapin

Don't confuse this plant with an oak that has a similar name. They do have leaves that look alike, but the oak tree will give you acorns where the chinkapin will have a spiny burr that houses one to four nuts inside. The Eastern part of the United States has a chinkapin that is native on the East coast from New Jersey down and west to Texas. If you go foraging for this nut, please wear gloves as these spines can become embedded in your skin. Don't get fooled by a red berry on this bush as this is a gall that was created by the chinkapin flower gall wasp. If you are lucky enough and find some burrs, you can dry these out a bit to make the nuts easier to get to. Don't ever expect to get a large number of nuts if you do find a chinkapin bush. Timing is everything, and you might have to search for the burrs at the correct developmental stage.

Clover

This plant can be found all over the world. Clover has been used as a cover crop all across the world. You can even find it in the Arctic and Antarctica. Clover is part of the pea family. Every part of the plant from its roots to its blossoms can be eaten. The blossoms are pleasant tasting but not the other parts. When you are harvesting blossoms, don't pick any brown ones. You want fresh ones, whether they are red, pink, or white, even though the white ones are the tastiest. You can put them in teas or roast them. Young leaves can be eaten raw but no more than about one cup. More mature leaves need to be cooked. Clover has beta carotene, protein, vitamin C, some B vitamins, bioflavonoids, inositol, choline, and biotin. Just a quick warning here: some people might be allergic to clover and don't know it. If you do decide to try it, only consume a small amount until you know for sure that you aren't allergic to it. NEVER ferment any part of this plant. You want to consume this plant either totally dried or totally fresh: nothing in between. Last but certainly not least, clover that is grown in warm climates can contain a small amount of cyanide.

Common Violet

Violets can be found in North America, Australia, and Europe. Violets are a native species. They are essential for pollinators like bees. There are other colors of violets like yellow, lavender, periwinkle, white, magenta, and of course purple. They are a tasty wildflower but are only around during early spring. It has a unique aroma and flavor that has been enjoyed for hundreds of years. They were used in baked goods and candies until WWI. When roads became more prevalent, the areas where they grew were torn up. Its most famous use was in Crème de Violette, which is made by infusing violets into natural spirits. The leaves can be put into salads. It has a flavor similar to lettuce and sweet peas. The flowers have a unique flavor that is delicately floral and not as intense as roses. They can be put into salads, but they will get lost in all the other flavors. If you have huge amounts of flowers, you could make jam or jelly, but the flowers are best when infused into alcohol or made into syrups.

Cranberry

Wild cranberries can be found all across the northeast part of the United States and Canada. They are similar to wild blueberries, and they like acidic soils. Most can be found growing in dense patches along the edges of small ponds, swamps, bogs, and lakes that are cold. They will be ready for harvesting around October and September. They can be eaten straight off the vine, or you can bag them and freeze them to be used later. If you are lucky enough to find some, look close to the ground as the fruit likes to hide among the foliage. When you get your fruits home, give them a good washing. Get rid of any broken or bad one and throw out all debris and stems. Shake off the excess water and place them into freezer bags and freeze. You could also turn them into juice or compote.

Dandelion

Dandelions are quickly becoming known as a superfood. These beautiful yellow plants can be found all over the world. They are higher in vitamins than spinach and kale. They are full of antioxidants and vitamins. Just a half-cup of greens contains more calcium than eight ounces of milk. They are also a good source of vitamins K, A, and C. They are rich in potassium, too. They can help stabilize blood sugar and makes them a great choice for diabetics. Dandelions can lower cholesterol, manages high blood pressure, slows aging, and detoxifies the liver. Every part of this plant is edible from the bright yellow flowers to its root.

Dayflower

The dayflower was introduced to North America from Asia and can be found prevalent in Florida. There are some native North American dayflowers in a different genus. It can be confusing. Most can be found growing peacefully alongside spiderwort. The young tips and shoots are great fried, steamed, boiled, or as an herb. The flowers can be eaten right off the plant or added to salads. The roots can be eaten but get slimy when cooked.

Devil's Walkingstick

The Devil's walkingstick is native to eastern North America. The fruits and leaves are edible. This plant is a silly looking plant. It is a part of the ginseng family but doesn't have the qualities of ginseng. It is great for toothaches and rheumatism. Native Americans and southern herbalists use the berries and inner bark as a pain reliever for inflamed gums and sore, arthritic joints. Eating a few berries raw is fine, but if you want to eat more, you should cook them and turn them into jelly. They have a bitter flavor. You can also infuse them in liquor to help with rheumatism. Cherokees have used the roots as a salve to help will skin issues like boils.

Dock

Most species of the dock are edible, but the broadleaf and curly dock are the most common in the United States and Europe. The dock plant has a tart flavor similar to lemons. If eaten in large quantities, it could cause kidney stones like spinach. When the leaves are young and tender, they will have a more tart flavor. From early to the middle of spring, you can pick young leaves to eat either raw or cooked. You can sauté or boil the leaves to get the most out of their flavor. They are great in cream cheese, egg dishes, stews, soups, and stir-fries. Since dock has a short harvest time, harvest as much as you can and blanch, then freeze them to be used later.

Dollar Weed

Dollar Weed is a member of the carrot family, and you only eat the leaves. It has a taste similar to celery and carrots. You can use it to add flavor to stocks. You could also add raw leaves to salads. It grows prevalent in zones three to 11 in the United States and can be a rather invasive plant.

Duckweed

Duckweed can be found all over the world. It has a flavor that is close to sweet cabbage. Duckweed grows in poor water. It contains niacin, vitamin B6, vitamin A, vitamin C, five percent fat, 44 percent carbs, and 20 percent protein. This is an extremely tiny plant that looks like a meal floating on top of the water. You need to boil the duckweed, change the water, and then put them in smoothies, etc.

Elderberry

Elderberries can be found in Europe and North America. Elderberries can be helpful for flu and cold symptoms. Those beautiful white flowers that the pollinators love will soon turn into beautiful dark berries waiting to be made into cold remedies. The unripe berries, stems, and leaves are toxic to humans. Only use the ripest of berries. Stay away from the red elderberries as these are also toxic. Raw berries are a bit poisonous and could cause diarrhea, vomiting, and nausea.

Evening Primrose

Evening primrose is a biennial plant that is native to the United States and Canada. It has been naturalized in Europe, New Zealand, Australia, South America, eastern Asia, and Russia. It loves full sun and is drought tolerant. You can harvest the roots during spring means you will get succulent, sweet, fleshy roots. This plant can grow fairly large, and some roots will look a bit like parsnips. Once the roots have been washed, you can eat them raw. You can gather roots later on, but it won't be as fleshy. The leaves can also be used just like any leafy green. The flowers have an amazing taste and can be put into salads. The seed pods can be roasted and enjoyed. The stems when young can be used. This edible plant contains vitamin B3, potassium, calcium, beta carotene, carbohydrates, and protein.

Fiddlehead Fern

Fiddlehead ferns can be found in the eastern parts of Canada and New England. These plants only grow in wet areas. You will need to look along the edges of swamps, streams, and rivers. Fiddlehead ferns start emerging during the middle or late April or into early June. Check their locations weekly, so you don't miss your opportunity to harvest this plant. They need to be picked while the tops are still coiled tightly. The short stem is also safe to eat. Once you get them home, wash them well, and store in the refrigerator. Never eat these raw. They need to be cooked. They can be used in dishes just like any other green vegetable.

Field Garlic

Field garlic can be found all across the world. These can easily be spotted growing in the middle of fields, meadows, and lawns. If you know what chives look like, you won't have any problems finding these edible wonders. If you think you might have found a patch of field garlic, pinch off some of its leaves and give it s sniff. If it doesn't smell like garlic, don't eat it. Every part of this plant can be eaten. If the ground around the plant is a bit hard, you might need to use a hand trowel to dig around the base of the plant. You can use this plant just like you would green onions or chives.

Forsythia

Forsythia can be found all over the world. The pretty yellow flowers can be eaten right off the plant, but some say it is a bit bitter. It has several medicinal properties like a skin tonic or diuretic. It was used in China to treat strep throat, bronchitis, and colds because it clears the body from toxic heat like dermatitis, skin eruptions, chills, fever, and sore throats. The flowers can be dried or steamed, used as infusions or decoctions, or turned into teas. You can also toss the flowers into your spring salad for a bit of color.

Garlic Mustard

Garlic mustard is a common weed in North America and Europe that seems to find new territory in the United States. By April or May, you should be seeing this plant popping up everywhere. To harvest this plant, just pull the entire plant up. You will be doing the plants around it a favor as this plant can choke out native plants. Your main focus needs to be how many and the size of the leaves. If possible, find plants that are tall and have large leaves at the base. These give you more to work with. The larger leaves are not as bitter. Try to get only the leaves as the stems can be a bit fibrous and stringy. If you won't be using the leaves quickly, place them in water and store in the refrigerator. They will keep for a few days. You can eat the leaves raw or cook them like any other green.

Glasswort

Glasswort grows in brackish waters and salt marshes. It can be found in North America, Europe, and Africa. It is ready for harvesting by late summer. They taste similar to asparagus and can be used either cooked or raw. The easiest way to cook them is by either steaming or sautéing. Don't overcook them as it will lose its flavor. Glasswort contains bioflavonoids, iodine, calcium, iron, vitamin C, B vitamins, and vitamin A.

Greenbrier

Greenbrier is native to the United States but can be found in Ontario, too. When harvesting this plant, you need to get it while it is very tender. Just snap the tendrils off wherever they will break off cleanly. You can eat them just like they are, or you can chop them up and put them in salads, soups, or stews.

Ground Cherry

Ground cherries are related to tomatillos and tomatoes. They can be found in most parts of the world. The fruit can be either cooked or eaten raw. You could also turn them into preserves or pies. The fruit normally falls off the plant before it has ripened. It will take another two weeks until the husk has dried out, and the yellow fruit emerges. You can gather the fallen husks and ripen them at home. The fruit will store longer if left in the husk. If the fruit is green, don't eat it as it isn't ripe. If any fruit tastes bitter, you need to cook it first. If it still tastes bitter, throw it out.

Ground Ivy

Ground ivy looks a lot like dead nettle or henbit. It is a native of southern Asia or Europe but was brought to North America around the 1670s for medicinal uses. Ground ivy can now be found all across North America except for Nevada, Arizona, and New Mexico. Young plants can be added to soups and stews where the older leaves can be used for medicinal purposes and teas.

Groundnut

Groundnut is a member of the bean family but looks like a potato. They are native to North America east of the Great Plains. You can eat the shoots, flowers, and beans just like you eat regular string beans. The tubers are what is eaten the most. They look just like any bean that you might see in a garden, and they smell like a bean. The plant also grows beans that can be cooked and eaten just like regular beans. The tubers are what you are after. They can be harvested at any time that the ground isn't frozen. They do taste similar to potatoes but a bit sweeter but not as sweet as a sweet potato.

Hawkweed

Hawkweed is part of the sunflower family. You can find this plant across North America and Europe. The young shoots, leaves, and roots are what you want to harvest. The leaves can be eaten cooked or raw, and the roots can be roasted as used as a substitute for coffee. Hawkweed is full of antioxidants and minerals. Once the flowers have appeared, the leaves can be mashed to soothe insect stings and bites. The leaves can be turned into a tincture to be used as an appetite stimulant, fever reducer, or cough suppressant.

Hawthorn

Hawthorn can be found across most of the world. There are over 1000 species of hawthorn in the United States alone. Hawthorn is full of micronutrients, minerals, nutrients, and other natural compounds. It is the oldest known medicinal herb found in records dating back to the first century. Its main use was to treat heart problems but can be used as a tonic, an anti-inflammatory, immune-booster, or for digestive problems. They taste a bit like apples and make great pie fillings, jellies, and jams. The leaves of this plant are also edible and can be harvested in middle to late spring. The berries will be ripe during early or late autumn. When they are ripe, you can strip them from their branches, just be careful of the thorns.

Hazelnut

Hazelnuts are full of fats and protein, as well as tons of flavors. The American hazel grows over almost all eastern North America. The beaked hazel grows from the lower half of Canada down to the northern part of the United States. You need to harvest the clusters while they are still green if you wait, most of the nuts will be on the ground. Hazelnuts can be used like any other nut.

Hickory

Most of the hickory nuts found in the United States will be edible. This most popular one is the pecan that is limited to the south. In New England, the most common is the pignut and shagbark. When you have harvested your hickory nuts, place them in the sun for a couple of weeks before trying to break into them. The meat will shrink away from the wall of the shell, making it easier to remove. You can use hickory nuts just like you would any other nut.

Honeysuckle

There are more than 180 species of honeysuckle. These can be easily found growing along roadsides and hedges all across the world. The flowers of honeysuckle can be harvested and added to ciders and liquors to make a tincture. Honeysuckle has delicate floral notes. It pairs well with rose, elderflower, strawberries, peaches, sage, mint, yuzu, and citrus. Try using honeysuckle syrup in recipes rather than honey in a "Bee's Knees cocktail" with rum, brandy, whiskey, or tequila.

Hornbeam

Hornbeam is native to the eastern part of the United States. It makes great firewood, but be careful not to overload your stove as it can overheat quickly. The wood from the hornbeam can be used to make tool handles and sleigh runners. The bark and wood are medicinal. Infusions or teas that are made from the bark can be used topically for pains, including baths, to help relieve arthritis or muscle pains. You could use it as a mouthwash to help toothaches.

Horsemint

Horsemint is a plant that you have probably never noticed until you have learned to recognize it. It grows from eastern Canada south to Florida and then west to Michigan, California, New Mexico, and into Mexico. It makes a nice tea but can be brewed as a stronger herbal medicine. If taken in large amounts, it could be fatal. Whether you use it as a weak tea, strong brew for flu symptoms, or a poultice for arthritis, it is a pretty plant to have in your garden.

Horsetail

Horsetail is a perennial that grows in the Middle East, Asia, Europe, and North America. It is also known as the scouring rush or puzzle plant. Horsetail can be used to help strengthen tissues like bones, nails, hair, and skin, mucus membranes, arteries, ligaments, cartilage, and teeth. Horsetail can reduce inflammation and help strengthens lung tissue. It can improve bladder and kidney health by helping the body become resistant to UTIs.

Indian Cucumber

Indian cucumber is a native plant to North America. The roots or tubers and the leaves are edible. It does taste a bit like cucumber. It isn't very common, so please don't take too much as you can eradicate the entire species. They taste great in salads because of their fresh flavors. If you would like to eat Indian cucumbers, dig some up and replant them in your garden. The roots have anti-convulsion and diuretic properties. They can be brewed to make tea to stop seizures in children. The crushed berries can be made into an infusion.

Japanese Knotweed

Japanese knotweed can be found anywhere in the world. If you do find a patch of this plant, get as much as you want. You won't kill this out. Try to harvest the plants during the middle of spring when they can be broken off easily. As they mature, they will be stringy, and you might have to peel them. The roots aren't edible, so don't bother with them. If there are any pieces of the knotweed that you don't use, make sure you boil or burn them as it can root from the tiniest of pieces. Knotweed can be eaten as a snack, in salads, or raw. You can pickle or cook them like asparagus.

Jerusalem Artichoke

Jerusalem artichoke is native to the central part of the United States. It is a wild sunflower and can grow up to 12 feet tall. The tubers have been used for food by the Native Americans long before we came from England. The carbs in this plant come from insulin rather than starch. Harvesting needs to be done after it has flowered and before it frosts. It would be best to leave them along for several years so the insulin can convert to fructose. Tubers will be easier to digest and sweeter. Eating them too early can cause a lot of gas.

Mark the plants while they are flowering and then use a shovel to dig up the roots after the stalks have died back.

Jewelweed

Jewelweed is a native plant of North America. It can be used as an herb or as an antidote for poison ivy, oak, and sumac. The light green, young stems can be sautéed with onion and garlic and added to dishes like stews and soups. You could also add it to red beans and rice. They taste a lot like collards but are tenderer. Jewelweed can be used as an antihistamine and anti-inflammatory. Mix ground up jewelweed with some petroleum jelly and put in on the skin before your trek through the woods.

Kousa Dogwood

This plant is native to Asia but has been brought to the United States for ornamental gardening. The kousa dogwood has a taste between a mango and an apricot. There aren't many recipes for this out there, but I turned some into preserves and jams. They are a bit bitter when cooked, but I figured this was because of the skins. I rip them open and push out the gooey center. Push the goo through a sieve to get rid of the seeds to keep bitterness at bay.

Kudzu

Kudzu is a very versatile plant that we just don't use enough. Kudzu flowers smell a lot like the grape-flavored gum that children love. You can smell it from hundreds of feet away. Kudzu was brought into the US during 1876 as a part of the country's centennial celebration. Japan built a garden using Kudzu in Philadelphia. You can eat kudzu in several different ways. Young leaves can be eaten as is or juiced. You can dry them and put them in teas. Shoots can be cooked and eaten like asparagus. The blossoms can be turned into jellies or pickles. The root is an edible starch. Old leaves can be fried and eaten like potato chips. When cooking with kudzu, just let your imagination guide you.

Lambs Quarters

This is also known as wild spinach. It can be found everywhere in North America. Wild spinach goes a long way in describing its texture and taste. It is best to harvest the younger leaves as the older ones get bitter. The flower buds and leaf tips are a good choice for foods, too. These can be used as a substitute for asparagus or broccoli. You can use them in smoothies, omelets, stir-fries, and salads.

Locust

This tree can be found throughout the eastern seaboard and the Midwest. The blooms of the black locust are just available for a couple of weeks in late spring. The blooms look a lot like pea flowers, but they hang in clusters like grapes. Everyone will be a creamy white color. The seed pods are poisonous. The leaves and bark are also toxic, so make sure you get rid of any when you harvest your flowers. The whole flower is edible. When you have finished your harvest, leave your bag open so the spiders you have captured can escape. They taste a lot like sweet peas. And the base is a bit crunchy like celery.

Lotus

Some people consider the American lotus to be more American than apple pie. American lotus has been the main food source for Native Americans and can be found south and east of the Rockies and in parts of California. The young seeds, flowers, shoots, and roots are edible; it was the roots that the Native Americans sought after. This plant gives you phosphorus, calcium, potassium, sodium, and some minerals. The seeds are low in fiber but are a great source of oil. Seeds that are about half ripe can be eaten cooked or raw and taste very similar to chestnuts.

Mahonia

This plant is native to the western part of the United States. Its berries can be used in confections, beverages, jams, jellies, and pies. They can be fermented to make wine. The flowers can be used to make a drink similar to lemonade or eaten as they are. The young leaves can be simmered and eaten as a snack.

Mallow

It is most common in the Bay Area of California. The young leaves can be eaten in salads even though they have a texture that is interesting and isn't very flavorful. You can use mallow leaves as an herb in cooking. You can dry the leaves and then grind them into a powder and then add them to smoothies, soups, or gumbo type of dishes. This powder can thicken the dish. The immature seedpods or "cheese wheels" are great as a snack or similar to okra if sautéed.

Maple

These majestic trees can be found throughout North America. Sugar maples are famous for their syrup, but most people don't know that other maples can be tapped for syrup, too. Native Americans would drink fresh sap as a refreshing drink. The inner bark of all maples can be either cooked or eaten raw. The young leaves and seeds are also edible.

May Apple

Mayapples can be found in the eastern half of the United States between Florida and Quebec. The fruit is the only part of the plant that is edible, but don't let the seeds go down with the fruit. It has a taste between guava and pineapple.

Milkweed

Milkweed can be found in the eastern part of the United States and the southern part of Canada. The young shoots, pods, and leaves can be eaten once they have been boiled. The big question is, how many times do you need to change the water, and does the water need to be boiling when you put the plant in? Some so-called experts say you shouldn't eat this plant at all while others say it's fine, so who do you believe? You experiment for yourself but stay safe. If after you have cooked the milkweed, it still has a bitter taste, don't eat it. Just give it a taste and wait about 30 seconds, your body will let you know. Native American tribes ate the immature pods, buds, and young sprouts. Flowers were dried to use during winter.

Miner's Lettuce

Miner's lettuce can be found all over California. It was eaten a lot by the miners during the gold rush. They consumed it to prevent scurvy. They learned this trick from the Native tribes in the area. Miner's lettuce is a great source of vitamin C, vitamin A, and iron. It has a delicate flavor and a crisp texture. You can eat it just like any other lettuce. You can harvest and eat it at any time, but it is best to get before it flowers.

Mountain Mint

In spite of its name, it isn't a true mint but is closer to the members of bee balm. There are about 20 species, and all are native to the Northeastern parts of North America. The thin-leaved species are the only one edible. The wide leaved species have high quantities of pulegone, which is toxic to the liver and repels insects. It is great to keep mosquitoes away but not great for the tummy. The flowers and leaves can be used in teas or potpourri. They also work great in bathwater.

Mulberry

Mulberry trees can be found in North America in zones five through nine. If you are ready for purple fingers, find a mulberry tree, and pick all the fruit your heart desires. You could place a cloth under the branches and shake the branch. The berries will fall off, and you have your harvest. Give them a couple of baths under cool water to get rid of any bugs or leaves. They don't keep well, so eat them quickly. They can be made into pies, jams, puddings, ice cream, and cocktails.

Nasturtium

You can enjoy the leaves and flowers of this pretty flower. Most people enjoy them raw and added in salads. They have a peppery taste. They are native to South and Central America, but they can be easily grown just about anywhere.

Nettles

The big thing to remember with nettles is that they sting, so wear gloves. Nettle grows pretty much everywhere, except Hawaii. They like stream banks and disturbed areas. They are better before they flower. You need to blanch or steam them to get rid of the sting, and then you can use them like any other green.

Oak

The oak tree makes acorns, and acorns are edible. These trees can be found in eastern North America. All acorns are edible. Place the acorns in a water bath. Get rid of any that float. Once soaked, dry them. Crack them open using a hammer and woodblock. Late September through October is the best time for acorns. Once shelled, you'll want to soak the acorn in several water baths to get rid of the tannins. You can dry it out again and make flour or nut butter with your acorn.

Parsnip

The wild parsnip can easily be confused with the wild hemlock, which is poisonous. You need to make sure it has a flower because that's when it can be more easily identified. You can cook a very small amount of the plant to test it, and if it tastes bad, don't eat it. If it tastes bad, then there could be toxins. Cook them just like you would carrots are parsnips.

Partridge Berry

The leaves and berries are the edible parts of the partridge berry. Leaves are often made in a tea. The berries can be used in any type of culinary dish, although they do tend to be bland. They can be found in rocky and sandy areas. They are also found around red maple swamps and bogs. It is most common in eastern North America.

Passion Vine

Also known as a maypop, passion vines can be found in the southeastern region of the US. They are fairly common. You can use the leaves, juice, ripe fruit, and flowers. They are often used to make tea, preserves, drinks, and the fruit can be eaten raw.

Pawpaw

Pawpaw trees are common in eastern North America, especially in the Appalachian region. Finding fruit in the tree is a little more luck of the draw. They are ready to be picked from August through October. For some, raw pawpaw's don't sit well on their stomach. They are great turned into ice cream or pudding.

Pear

Wild pears aren't as easy to find as apple trees are, but they're out there. They look just like the pears you are used to and can be found in the same areas as apple trees. You can eat them like it is, or you can cook with them just like you would any pear.

Pecan

Pecans, like most nuts, have an outer hull that has to be removed and then a shell you have to get rid of before getting to the nut. Pecan trees are common all over Texas and in other similar climates. The hull will naturally be shed later in the season, but make sure you pick the nut off the ground before the wildlife has a chance to grab it. You can use the pecans just like you would use the ones you buy from the store.

Peppergrass

This is a native species in the mustard family. Its flavor tends to sneak up on you. The seedpod is the best part of the plant. They have an arugula-like flavor. You can dry out the peppergrass for winter use. You can use it just like any other seasoning. It is native to most of North America.

Persimmon

The wild persimmon is one of the last fruits ready to be picked in late fall and early winter. Persimmon trees grow in full or partial sun, and they like infertile soils. Fruit should be harvested when they start to wrinkle. You can enjoy them raw. You can also use it in desserts and preserves just like you would with other fruits.

Pickerel Weed

This plant is found in eastern North America as well in Argentina. It is also in Oregon. The seeds can be cooked or eaten raw. It can also be boiled and eaten like rice. You can also dry it and turn it into a flower.

Pine

Pine nuts come from pine, or pinon, trees. They are expensive to buy because they are hard to gather. Pine needles can also be eaten. They can make a great tea, or mixed into a recipe for a spicy kick. Pine trees can be found in most areas o the US, specifically in coniferous forests.

Pipsissewa

This oddly named plant can be found in almost every state except for the central southern and Midwest states, as well as Florida and South Carolina. It prefers the dry woods. Only the leaves of the plant are used, but they are tough and unpalatable. They are best used to make a tea.

Plantain

The young leaves can be eaten raw in a salad, or you can cook it. They are full of vitamin B1 and riboflavin. It prefers sunny areas and grows in most areas in the US.

Pokeweed

First, be careful with pokeweed because it can be toxic if misused. Young shoots can be boiled in two changes of water and will taste like asparagus. Poke berries can be cooked, and the liquid can be used to color canned veggies and fruits. Do not eat the berries; they are poisonous raw. It commonly found in central and northern North America. It found in rich soils.

Prickly Pear

Prickly pear is a cactus, so it is only found in desert climates. You can find them in the store labeled as nopales. If foraging for them, make sure you have gloves and carefully remove the spines and glochids, which are tiny hairs that are hard to get out of the skin. You can roast them, or eat them raw. Just make sure you get rid of the glochids.

Purple Dead Nettle

These are only available during early spring, so make sure you grab some while you can. It is a mint and is commonly confused for ground ivy and henbit, but of which are edible. It is easy to spot with its purple tops and grows just about anywhere, especially in your yard. It is commonly dried and used to make tea.

Purslane

Purslane grows everywhere fairly easily. It is considered an invasive weed in most of North America, so it is a sustainable plant to forage. There is a deadly lookalike, the potentially deadly spurges. The main thing to remember is that purslane will not have a milky sap. You can use purslane just like any other green.

Queen Anne's Lace

Also known as the wild carrot, is biannual. The first year it sends up the flower stalk and goes to seed in the second. You can harvest the roots during the spring or fall of the first year. The wild carrot can be used just like the garden carrot.

Ramps

These are some of the earliest wild edibles you can find. They will occur in higher elevations in eastern North America from Canada to Georgia. You can easily spot them by their broad leaves. Ramps can be found under the dense deciduous forest canopy. They can be used just like any other onion, but keep in mind, they are stronger.

Raspberry

Raspberries will start to ripen during the first part of the summer. They can be found all over the US. You can use the leaves and berries. The raspberry leaves are particularly helpful to women and can be made into a tea.

Redbud

The Eastern redbud tree is native to Eastern North America. In the spring, the flowers become a bright reddish-purple that can easily be seen. You can harvest the flowers and use them to make syrup, jelly, and cakes.

Rivercane

This is an uncommon plant, so forage wisely. You can use the seeds and young shoots and should be cooked or steamed. They can be found at river banks and are found throughout most of the south and eastern parts of the US.

Rose Hips

Rose hips start forming in late summer and through fall. Rose grows wild in most places, so you should be able to find the hips. Hips only form where the flower was, so if you have your own roses, don't cut them. Rose hips can be used as a fruit and dried to make tea.

Sassafras

This is a deciduous tree found in eastern North America. It has been used for many years to cook with, but be careful because it does contain safrole, which is toxic. The bark can be made into a tea. The root is most commonly used and was used to make the first root beer.

Serviceberry

These are native to North America. They taste like a cross between a grape and blueberry. They are easy to spot because there aren't any lookalikes. The smooth gray bark and have a showy, star-shaped white flower. They can be used as a substitute for blueberries.

Shepherd's Purse

This green shows up in early spring. Before the flower stalks appear, the leaves are great in salads or cooked like greens. They are in the mustard family. They often grow in grain fields, lawns, and gardens. The leaves can be a cabbage and cress substitute.

Smartweed

True smartweed will be very peppery. If it isn't, then it isn't smartweed. It is common throughout North America. It is best to soak the leaves in several water changes before cooking it. They can be added to salads or used just like you would other greens.

Sorrel

Sorrel is normally found in your lawn in most areas in North America. All you need to do is pull some of the upper parts of the plant off, or strip away some of the uppermost leaves. You can cook up sorrel just any other green.

Spicebush

The leaves, twigs, and berries are edible. The spicebush is commonly found throughout most of eastern North America. It is an understory shrub of moist forests and likes medium and light shade and rich soil. The twigs and leaves are often made into a tea. Only the female plant will produce bright red berries. The berries aren't used much because they don't grow in abundance, and they tend to rancid due to their fat content.

Spring Beauty

Spring beauty is part of the same family as purslane and can be used in a similar way. It is commonly found in the eastern 2/3's of North America. The plant prefers a dappling of the sun in the spring, and moist to slightly dry conditions. It can take a while to collect spring beauties due to their small size.

Sumac

The young shoots and roots can be peeled and then eaten raw. You can also eat the fruit raw, cooked, or made into a lemonade-like drink. It is a deciduous shrub native to North America and can be found in all mainland states and in southern Canada. It can be found in thickets, roadsides, and open fields. It tends to be invasive.

Thimbleberry

The thimbleberry belongs to the same family as blackberries and raspberries. While botanists would not call them berries, and refer to them as drupelets, we will simply call them berries. You normally can't find a whole bunch of them, and the bushes are normally small, but you can find enough to make some jam. They can be found in the same areas as raspberries and blackberries.

Thistle

Most places have a thistle plant, but the most common is bull thistle. They like areas with plenty of sun and little traffic and can grow to five feet tall. You can dig up the roots and cook with them. If you don't like the taste of burdock root, you may prefer thistle root. All of the thorns have to be removed from the leaves and the plant before you can cook with them. Most people find that the leaves are worth the work of removing the thorns. Make sure you think about using eye protection, as the thorns can cause a lot of damage. Remove all of the leaves from the stalk and then rack the thorns off of the stalk. These are good added to a soup. The flower can also be used and as an artichoke-like heart in the center.

Watercress

This wild plant is full of vitamins and minerals. It has a mildly hot must flavor, and is great on sandwiches or in salads. You can take the seeds and grind them into a powder to make mustard. It is often found near springs and brooks.

Wild Garlic

Wild garlic can easily be found by its distinctive scent, so you can't confuse it with anything else. It is best to harvest just the greens and leave the bulbs so that they can come back the next year. You can use the greens to make a tasty pesto. You can typically find some type of wild garlic in every part of North America.

Wild Grape

The wild grape is a vine, so it does not have a solid, upright trunk. It is a climbing vine and can completely envelop trees and bushes. It can grow in many different areas and all over the world. The grapes start as tiny white flowers and develop into green grapes. You can enjoy the grapes as soon as they form, but they taste better after the first frost. They are great in salad, or you can make your own jelly with them.

Wild Lettuce

This is a common plant in North America and is often viewed as a noxious weed in cracks in sidewalks, fields, and in yards. It is normally tall and had dandelion-like leaves. Some types will have yellow to reddish-yellow flowers. Since it is typically prickly, it is best if you cook it slightly first, or at least blanch it. Otherwise, you can use it similar to how you would use regular lettuce.

Wild Onion

Wild onions or ramps are some of the best wild foods during the springtime. Ramps have large wide leaves, and fairly easy to spot, especially in Eastern woodlands where they sometimes carpet the forest floor for acres. Some wild onions aren't as easy to spot. You can find some type of wild onion in any area of the US. They all send up grass-like shoots. If you see grass that looks like an onion, pinch a piece and sniff your fingers. If it smells like onion or garlic, then it is like a wild onion. There are plenty of poisonous plants that look like onions, but none of them will smell like one as well. They can be used just as you would use any type of onion.

Wineberry

Wineberry is one of the most abundant brambleberries. It is often placed on the invasive species list, and luckily they are also tasty. You can look for the wineberry around the same time as the blackberry season. They prefer full to partial shade and can be found in clearings, edges of fields, and in parks. They are orange-red and look similar to a raspberry. You will get sticky while picking them. You can enjoy them fresh, but they also make delicious baked goods and preserves. You can also make wine from them.

Yarrow

Yarrow is commonly used as a tea and is often taken before bedtime because it can help with insomnia. Dried, it can be used to flavor any dish that you would like and is similar in flavor to sage. It is commonly found in meadows, pastures, and old fields in eastern and central US and Canada.

Yaupon Holly

Yaupon holly leaves are normally dried and used for tea. You can find them all year, and they contain antioxidants and caffeine. The berries should be avoided because they can make you sick. Make sure you don't confuse it for the toxic Chinese privet. It is common in the southeastern US.

CHAPTER 4

Compendium Of Medicinal Plants

Many of the edible plants are also medicinal and can be used to help treat common ailments. They should not, however, replace modern medicine, especially for medical emergencies. When you are out foraging, keep an eye out for these medicinal plants to take home and enjoy.

Agrimony

This is a Chinese herb that is commonly used to stop bleeding. It is not as commonly used today but has its place in herbalism. It is safe in moderation for healthy people. It has neuroprotective properties, anti-inflammatory, and antioxidant. Agrimony contains tannin and volatile essential oil. It has an astringent action and is commonly used as a mouthwash ingredient. It can be applied externally to heal ulcers and sores. It grows in the US, Canada, and Europe. Its natural habitat is fields and woods. It has one to two-foot branched stems covered with silky, fine down and yellow flowers.

Aletris Farinosa

This is an uncommon wildflower with grass-like leaves. It is also known as ague root, true unicorn root, colic root, and star grass. It has commonly been used to help with female-specific health complaints. It can become toxic if too much is ingested. Since it is rare, it likes to be a protected plant in most places because cultivation is difficult.

Angelica

Angelica is commonly used in herbal medicine. It is a great tonic for children and women, as well as the elderly. It is believed to help strengthen the heart. Powdered angelica root is believed to cause disgust for liquor. It can help prevent the growth of bacteria. It can also be used for obstructed menses and shouldn't be taken in large amounts by women who are pregnant. An infusion can be used as a gargle for sore throats. It can be used as a wash for the face to prevent acne. The fresh angelica root is not edible and can be poisonous. Angelica contains glucose, fructose, iron, magnesium, potassium, riboflavin, sucrose, thiamine, zinc, and vitamin B13. It is a biennial native to Eastern North America. It can be found in thickets, moist cool woodlands, bottom-lands, and shady roadsides.

Black Cohosh

This plant is a giant member of the buttercup family and is native to North America. Chinese and Native American herbalists used this plant for many different ailments. Women will often use black cohos to help treat the symptoms of menopause, and it can help with cramps and induce labor. Be careful; in large doses, black cohosh can damage the liver.

Bloodroot

Bloodroot is commonly used in herbal medicine in small doses for throat infections and bronchial complaints. It is commonly used in different pharmaceuticals, mixed with other items to help treat heart problems, dental uses, and to help with migraines. A paste made from bloodroot can help to treat tumors, warts, and other skin diseases. You can even find bloodroot salve for this purpose. It is most commonly used externally, but when taken internally, you have to be careful because it has opium-like alkaloids, and an overdose is possible. It is native to North America and can be found in moist woodlands.

Blue Cohosh

This attractive woodland herb is become endangered because of over-harvesting. It rarely grows more than two and a half feet. In Tennessee, it will bloom in early April and is commonly found on wooded slopes. It has historically been used as a uterine tonic and to help with difficult labor. Its seeds can be used as a substitute for coffee. The berries are poisonous, and any part of the plant you want to use for medicinal purposes, do so under the advice of a qualified herbalist and medical practitioner. Pregnant women should not consume.

Butterfly Weed

This is an edible plant that is commonly used by herbalists. Also known as pleurisy root, it can be used as a vasodilator, tonic, expectorant, diuretic, diaphoretic, carminative, and antispasmodic. The root has been used to help treat rheumatism, dysentery, and diarrhea. It is a valuable herb for anybody how has lung problems. A poultice of the plant can be used to treat skin ulcers, wounds, bruises, and swellings. Too much butterfly weed can become an emetic. The seed pods are edible. Cook them when they are young and harvest them before the seed floss forms. The flowers are also edible when cooked. It can be found in Eastern North America. It is often found in dry open fields and grassy places.

California Poppy

The California poppy does not have any opium in it, nor does it have any addictive qualities. However, it can be used to treat similar problems that regular poppy can help. It has milder tranquilizing and pain-relieving problems. Instead of creating psychological problems that opium poppies can cause, it is mentally stabilizing. It is normally used as a tincture since the infusion is often bitter. The above-ground parts are normally harvested during the flowering period. Make sure you don't use California poppy before driving or any similar tasks until you are familiar with what it can do. It tolerates dry, hot conditions very well. It is common in California and is the state flower.

Chamomile

This is one of the most common tea ingredients and can easily be found in the grocery store. It has a mild sedative effect and is an effective treatment for insomnia and other nervous conditions. There is a small risk of allergy, but it is generally considered one of the safest herbs. It can also be sued to relieve gas and menstrual cramp. It can also be used as a mild laxative. It can also be used to help with allergies, the flu, and colds. It is also commonly mixed into lotions and soaps to help soothe irritated skin. It is effective for softening the skin, and relaxing tired, achy muscles.

Cleavers

A word to the wise, fresh cleavers plant can cause contact dermatitis for some. You should wear long sleeves and gloves when harvesting. It has been used to help cleanse the lymph and blood system. It is considered a diuretic, so make sure you stay hydrated while using this herb. It should be cooked down or processed in some ways before using it. Cleavers can be found worldwide, and nobody really knows where it originally came from.

Echinacea

Also known as purple coneflower, it has been used for many years in herbalism to help support the immune system. It is a popular herb taken during the cold and flu season. Echinacea extracts have been found to improve the cellular immune function in healthy people and in patients with AIDS. It doesn't fight off bacteria but makes the immune cells more efficient in fighting off the bad cells. It can be found in most states in the eastern US.

Ginkgo Biloba

The Chinese have been using this plant for well over 5000 years. It has been linked to alternative medicines for asthma, Alzheimer's, kidney disorders, heart disease, and to build energy. It is an adaptogen, which can help the body out during stressful situations.

It is a perennial deciduous tree native to China. Ginkgo Biloba is one of the oldest species of trees on earth and dates back more than 300 million years.

Ginseng

Ginseng has a long history of herbalism that dates back to 5000 years and appears on several continents. It has been extensively used in Native American medicine. It is another adaptogen and can help the body adapt to emotional and mental stress, cold, heat, fatigue, and even hunger. Ginseng is often one of the ingredients in energy drinks, but it is not actually safe to combine ginseng with caffeine as it will accelerate the caffeine's effect on your body and can end up causing diarrhea. It is a perennial herb in eastern North America. It likes rich soils and cool woods, but check your local laws for when you can harvest ginseng.

Goldenseal

Goldenseal is actually endangered in the wild, so you should try to cultivate it on your own and not harvest it. It can be used for short periods of time as an antibacterial, antiseptic, and antispasmodic. It is often made into a tea for upset stomachs and other digestive issues. It can also be made into a mouthwash for mouth ulcers, pyorrhea, or sore gums. It is native to eastern North America.

Joe Pye Weed

Historically, this plant has been used to treat fluid retention, gallstones, and rheumatism. The roots are the most potent. The flowers and roots are commonly used to make tea to relieve urinary and kidney problems. It can also be used to induce sweating and break a fever. It is come in eastern and southern parts of North America and prefers moist meadows and woods.

Lemon Balm

This is a medicinal and edible herb that can be added to salads or other dishes. The entire plant can be used, fresh or dried, to make warm relaxing teas and refreshing drinks. It contains volatile oil citral and citronella, which can help calm the nerves, relieve insomnia, menstrual cramps, and depression, and can relieve colic in babies. It is common throughout Europe and is mainly cultivated in the US. It can be found in sunny fields and is easily cultivated by cutting or seed.

Lousewort

Also known as wood betony, it is an edible and medicinal herb that was commonly used by Native Americans. It is prized for its aphrodisiac and medicinal qualities. You can apply a poultice of the herb to sore muscles, tumors, and varicose veins. A tea can be used to treat bronchitis, cough, tonsillitis, and a sore throat. It is native to eastern North America and can be found from Nova Scotia to Manitoba, and south to Northern Mexico and east to Florida. It likes moist thickets and woods.

Motherwort

This has become known as a woman's herb, which can be helpful in all stages of life. It can be used to treat heart palpitations, nervous pain, and menstrual cramps. It can also be helpful in lowering blood pressure and has hypotensive properties. It can be easily grown from seeds and reseeds itself profusely. Motherwort likes moist ground and blooms in midsummer.

Mugwort

Mugwort has been used for years in herbalism. It has antispasmodic, antibacterial, nervine, purgative, diaphoretic, tonic, diuretic, and digestive properties. A tea made of the flowers and leaves can help matters of the digestive system. It is native to Europe, Asia, and Africa, and has been naturalized in many areas of the world. It can be found growing around hedgebanks and waysides.

Mullein

In many countries throughout the world, mullein is a proven medicinal herb and has been back by scientific evidence. It can be used as a sedative, hypnotic, fungicide, estrogenic, cardio-depressant, bacteriostatic, antiviral, antioxidant, anticancer, anti-inflammatory, antihistaminic, and analgesic. It can be found all over Europe, in temperate areas of Asia, and in North America. It can be found on hedgebanks, and most often on chalk, sand, or gravel areas. It likes dry, uncultivated soil.

Soapwort

This herb has been used is the time of Dioscorides. It can be used as a tonic, purgative, expectorant, mild diuretic, diaphoretic, depurative, cholagogue, and antiscrophylatic. It can be applied to the skin to help with itchy skin. You can also obtain a soap by boiling the whole plant, especially the roots, in water. It is great for delicate fabrics that can be hurt by synthetic soaps. Do not take in excess as the plant can destroy your red blood cells.

It is a perennial in Europe and has been naturalized in the US. It can be found in moist ditches, meadows, and near old home sites.

Spearmint

This edible herb can be used cooked or raw. It has a very strong flavor and is commonly used in teas, which leave the body feeling clean. It can also be used to make mint jelly. The essential oils are commonly used to flavor candy, gum, drinks, and ice cream. It is great for using as an antiseptic, to ease stomachaches, and as a diuretic. It is found in Central Europe and has been naturalized in the US and Canada. It likes damp, sunny places.

CHAPTER 5

Compendium Of Poisonous Plants

The most important thing to be aware of when you are foraging is poisonous plants. Foraging should be something fun, and possibly getting sick or dying from foods you have foraged isn't fun. Let's go through some poisonous plants that you should avoid at all costs.

Angel's Trumpet

This is a distinctive woody bush with trumpet-shaped flowers that hand like bells. They are most fragrant at night. It is a common ornamental plant in the US, so you probably won't find it in the wild. While beautiful, they can kill you if you eat it.

Castor Bean

The castor bean contains ricin, which is the most toxic naturally-occurring substances. The oil is often used medicinally, but only after the ricin is removed through processing. If eaten raw, every part of the castor bean seed can be fatal. It is an ornamental woody shrub with star-shaped glossy green leaves and feathery flowers. It is native to Africa but was introduce to North America. It can be found in the Eastern and Southern parts of the US, often in disturbed areas.

Corn Cockle

This was originally found in Europe but became established in the US. It used to be used in herbal remedies, but its toxicity makes it dangerous to us. It is covered in fine hairs. Every stem had a single deep pink to purple flower and a ribbed bulb. It is a hardy plant and can be found in disturbed areas.

Daffodil

This flower may let you know that spring has arrived, but it is a common cause of poising, especially for pets. Eating any sections of this plant can create stomach problems. They can be found throughout North American in many different habitats.

Deadly Nightshade

This plant produces a cherry-like fruit that turns a shiny black color once ripe. The ground plant contains light-purple bell-shaped flowers and pale green leaves. It is pretty rare, but it can be found in the eastern and southern US.

Death Camas

Early settlers found out the hard way that this plant was deadly. Believing that it was onion or the edible camas plant, pioneers were surprised when they become ill after eating it. It has white, star-like flowers that are clustered around the end of a spike-like stem. Its leaves are long and grass-like and look like an onion plant. It is native to western North America and can be found in meadows.

Foxglove

Another European flower that found its way to the US, foxglove, can be spotted by its flowers that grow in a cluster around a long stem. The flower is funnel-shaped and typically points down. They come in a variety of colors. You can find its grey-green leaves near the base of the plant and can be a foot wide. Every part of the plant is poisonous.

Giant Hogweed

As you could guess by the name, the giant hogweed can reach 14 feet tall and five feet wide. It creates a white umbrella-shape flower cluster that can have up to 50 rays on each cluster. It has lobed leaves and measures up to five feet wide. Contact with the sap can cause burns and blisters. It can be found in Maine, New Hampshire, Vermont, Virginia, Michigan, Washington, Oregon, Maryland, Ohio, Pennsylvania, and New York.

Iris

You can easily spot an iris from the showy flowers that have three large sepals and three interior petals that droop down. All species are toxic, and the toxin is more concentrated in the roots. It can be found all over the US.

Jack-in-the-Pulpit

After you know what to look for, you won't be able to miss this flower. The flower will only be three to four inches tall with a tubular base and hood that curves over the tube. The hood will normally be green with purple and white stripes. It will have several big showy leaves. It can be found in moist woody areas in North America. It contains a calcium oxalate toxin that is mainly in its roots. You can touch, but don't consume it.

Jimson Weed

Also known as devil's snare is another nightshade plant that is well known for its various poisonous family members. It can grow between two to five feet with a thick stem and trumpet-shaped purple or white flowers. It has irregularly lobed leaves and is stinky when crushed. The seed pods are just as wicked with spikes on the outside. Every part is toxic and can cause hallucinations.

Larkspur

This plant is part of the buttercup family, but it isn't your regular buttercup. It grows in spikes and can reach five feet tall. It has a raceme of flowers that begin close to the base and extends to the top. It had a hollow stem, which distinguishes from monkshood, which is also poisonous. Every flower has a dolphin-like appearance. It is mainly found in the western and southern US. It likes moist soil.

Manchineel

The manchineel has been named the most dangerous tree in the world because of all of the toxins that are found in nearly every part of the tree. The sap can create blisters on the skin and can blind you if it gets into your eyes. A simple bit from its fruit can be fatal, and the smoke from burning leaves and branches can hurt your lung and eyes. This is a tropical tree and is native to Florida.

Monkshood

While monkshood is poisonous, it also has an unpleasant taste, so accidental poisoning is very rare. You can spot the plant by its distinctive purple or white flowers. They are arranged in a spike-like cluster with a hood shape that inspired its name. It can be found in mountainous areas.

Mountain Laurel

During the mid-summer, this can be found all over the Appalachian mountains. The flowers are showy with petals that are shaped like a bowl with distinct purple marks.

Oleander

This is a small, hardy shrub that has leather, slender, long leaves. When in bloom, it has a funnel-shaped flower and is commonly planted by the roadside. The flowers grow single or pairs and are typically bright in color. All parts of very toxic and should not be ingested. It can be found on the west coast and in southern states.

Poison Ivy and Oak

These, along with poison sumac, are the most common poisonous plants that everybody had heard of. It is most know for causing itching, swelling, and rash when it comes in contact with your skin. If you were to consume this, just imagine what your insides are going to feel like. Leaves of three are something you should remember to avoid coming in contact with poison ivy. The leaves are reddish in the spring, green in the summer, and yellow or orange in the fall. It can be found pretty much everywhere except for Hawaii and Alaska, and certain deserts.

Poison Sumac

Poison sumac looks different from the ivy and oak. It usually grows as a tree, five to 20 feet high, and likes swampy areas. It has a red stem and multiple leaves that have a smooth edge. It is most commonly found in the southern and eastern US.

Rosary Pea

These are native to India and were introduced to the US, where it is now seen as an invasive weed. It can be found in pastures, roadsides, abandoned farms, and other disturbed areas. They are highly toxic, and a single seed has the potential to kill you. It has a distinctive red and black pea that is uniform in size.

Water Hemlock

This is sometimes referred to as poison hemlock and is part of the same family as wild parsnip. It is not, however, related to the hemlock tree. It can easily be consumed with Queen Anne's lace and yarrow, which are edible. Water hemlock can be found throughout the US. The difference between this and Queen Anne's lace is that the stem is hairless, and they may contain a purplish splotch. On Queen Anne's lace, the stem is hairy, and it will often have a single dark purple flower in the middle of each umbel.

White Baneberry

This plant is hard to miss. It is white berries on a red stalk. It's a good thing that the berries look quite creepy because eating them can cause cardiac arrest. It is found in eastern North America.

White Hellebore

This plant is historically known for being poisonous, particularly in ancient Rome and Greece. Its cup-shaped flower with egg-shaped leaves is all toxic. The sap and seed can leave nasty chemical burns on the skin. It is native to Asia and Europe but is often planted in gardens in North America.

White Snakeroot

This is part of the Aster family, which includes daisies. It can grow to three feet tall with thin stems and egg-shaped leaves with a toothed edge and pointed tip. Their flowers are clustered with tiny, hairy protrusions. It can be found in the eastern part of North America, where it grows in wooded pastures and in forests. They contain tremetol, which is very poisonous.

Wild Parsnip

This is part of the parsley family that includes dill, celery, and carrot. It can grow up to five feet tall and will have yellow flowers that create an umbrella-shaped cluster. It can be found by the roadside, in pastures, and in fields. The sap of the wild parsnip causes the skin to become more sensitive to light. Most people don't realize when they have come in contact with wild parsnip until they break out in a blistering rash after being in the sun. It can be found throughout North America.

Wild Poinsettia

This is also called Fire on the Mountain; it has lobed leaves with irregular red blotches around the base of the topmost leaves. It has a milky sap that helps you identify it. It is toxic and irritating to the skin. They are native to tropical climates in South America, but it can be found in the southern part of the US. These are commonly seen around the holidays.

CHAPTER 6

The Basics Of Foraging

It is important that you understand the basics of ethical foraging when you make your first foraging trip. While we will go over most of what you will need to know, you can find more information in the book on foraging. Ethical foraging means that you won't ever take so much that you deplete the area of certain plants. This can prevent you from being able to return back to this area to forage the next year.

The responsible forager will always have conservation of the plants in mind when foraging. There are some plants that you won't have to worry about, but this will all depend on your state. Plus, there are some plants that are invasive, so it can be helpful to the environment to take as much as you can. With that in mind, let's take a look at some ways to make sure that you are foraging responsibly and ethically.

Make Sure You Can Be on the Land

This should be something people should naturally be conscious of, but sadly it's not. You should not forage for edible plants on private property unless you have the property owner's permission. When you ask for permission, you have to make sure that the property owner knows that you are going to respect their land, take only what you need, and offer them something in return.

When you are on public lands, like national or state forests and national parts, make sure that you check the rules of each place because the regulations can be very different. Some places will require that you get a permit, while others don't. Some may limit how much you can harvest. There are some parks where you can forage for berries, but you can for mushrooms. Always make sure you double-check and follow their rules when it comes to foraging.

Know How to Identify Plants and Forage Safely

The quickest way to get sick is to eat something you think is edible but isn't. You have to make sure that you know how to identify wild plants before harvesting them. This will ensure that you don't end up eating poisonous plants, but it is also important that you don't end up taking something that can't be used. You should never forage something unless you are 100% positive of what it is.

If you choose to do spore prints of mushrooms to help with identification, harvest just one of them before returning to harvest more once you find out what they are. It is always a good idea to leave something untouched and growing for sustainability.

Safety should always come first, no matter what. You should make sure that somebody else knows where you are going before you head out and when to expect you to be back. This will ensure somebody will know to come looking for you should you get lost. Make sure you are dressed for the occasion as well, and that you have the right equipment with you.

Remember the Four Rs

The four Rs are roadways, right of ways, residences, and railroads. Most experienced foragers will tell you to avoid busy roads, right of ways, railroads, and any built-up areas because the edibles in these spaces may end up being contaminated by herbicides, particulates, fertilizers, and so on. But then there are some who believe that it all depends on what foods you are looking for. Generally speaking, plants adapt to the environment that they are in through their roots and leaves. This means the fruits, as long as they are at a distance from the ground, are typically fit and safe to consume. No matter what you are gathering, though, make sure that you use your common sense. If something arouses your suspicion or it seems to be contaminated, don't eat it.

Know Your Protect Species

Protected species lists will change all the time. To make sure that you don't hurt the environment or accrue costly fines, talk to a local expert about protected species before you head out foraging. However, if a plant has been labeled protected, then it is rare and nearly impossible to find. This means it is probably very unlikely for a forager, expert or novice, to be able to find decent amounts of any protected plant that is on the cusp of going extinct.

Don't Take the Only Plant

Follow the rule of abundance. If you can only find one plant is a sizable area, leave it untouched. You should never take the only edible plant of a wild species as it will likely be unable to regenerate. Remember that the planet, forest, and animals need that single specimen more so than you do. Similar to this idea, never take everything from a single spot. You should only remove less than 10% of any one plant.

Take Only What You Need

Forage only what you are going to need to make whatever you have in mind. Only take enough leaves to dry them out for the winter's tea drinking. Leave all of the others for the earth. You don't need to hoard more than your home is going to need.

Harvest Your Plants Wisely

You should harvest from plants that look like they have been stressed from fire, flooding, drought, or any other situation. Take only from healthy plants that are abundantly available. When you take part of the plant, harvest just the top 2/3's of the plant, leaving all the rest for it to regenerate and spread as nature would like it to. If you need the plant's roots, dig it up carefully and cut off parts of it with a knife instead of just ripping out of the earth. When you take responsible care during harvest, it will help to make sure the plant stays healthy.

As you make your way out into the world looking for wild edible plants, enjoy the beauty of nature. Allow the different plants to show themselves to you. When finding these edible plants, follow the steps that we have discussed to keep your foraging fun and safe for you and the earth.

CHAPTER 7

Recipes For Edible Wild Plants

<u>Buffalo Milkweed Pods</u>

You'll Need:

- Favorite hot wing sauce
- Water, .5 c
- Almond milk, .5 c
- Egg

- Turmeric, cayenne, oregano, and paprika, 1 tsp each
- Garlic powder, 1 tbsp
- Flour, .25 c
- Panko, 1.5 c
- Milkweed pods

You'll Do:

1. Start by getting your oven to 350.
2. Combine all of your dry ingredients together. Beat together the water, almond milk, and egg together and then mix it into the dry ingredients. Combine well.
3. Dip the milkweed pods into the batter and then lay them out on a baking sheet that has been lined with parchment. Let them bake for 15 to 20 minutes.
4. Once they are crisp, move them to a bowl. Pour in your favorite wing sauce and make sure that they are well coated. Spread them back out on the baking sheet and cook them for another ten minutes. Enjoy.

Cattail Rice

You'll Need:

- Pepper, 1 tsp
- Vegetable broth, 2 c
- Garlic salt, .5 tsp
- Uncooked rice, 1 c
- Diced onion
- Chopped young cattail shoots, 1 c
- Butter, 2 tbsp + 1 tbsp

You'll Do:

1. Start by adding the two tablespoons of butter to a pan and letting it melt. Add in the cattail shoots and cook until they are soft. This will take about five to ten minutes. If you need to, you can add in the extra better. Once soft, turn the heat down to low and add in the onions, pepper, and garlic, cooking until translucent.
2. As your veggies are cooking, cook your rice according to the package directions. If you need to, adjust the rice to water ratio according to the directions.
3. After the rice has been cooked, mix in the cooked vegetables. Cover the pot and let everything sit for two to three minutes. Enjoy.

Dandy Pasta

You'll Need:

- Favorite spice mixture, 2 tbsp
- Coconut oil, 3 tbsp
- Butter, .25 c
- Garlic, 2 cloves
- Red onion
- Dandelions with roots, 2 to 3 c
- Bowtie pasta, 2 c

You'll Do:

1. Start by cutting the roots off of the leaves right at the top so that the leaves will stay together. Finely chop up the roots along with the garlic and onion.
2. As this cooks, follow the directions on the packaging for your pasta and cook the pasta until done.
3. Melt your butter in a pan along with the coconut oil. Add in the spices and mix well. Add the dandelion root, onion, and garlic, and let this cook for five to ten minutes.
4. After it is done cooking, add in the cooked pasta and stir everything together. Let this simmer together for one to two minutes. Mix in the dandelion leaves. Let it simmer for a minute and enjoy.

Garlic Mustard Stuffed Mushrooms

You'll Need:

- Melted butter, 1 tbsp
- Grated cheese, 1.5 c
- Finely chopped onion
- Finely chopped lambs quarters, .75 c
- Finely chopped garlic mustard, 1 c
- Cream cheese, 4 oz
- White mushrooms, 20

You'll Do:

1. Start by getting your oven to 350.
2. Wash the mushrooms and get rid of the stems. Hollow out the insides of the mushrooms to make space for the filling. Cover your mushrooms as you work on the filling so that they don't dry out.
3. Add the onions, both greens, and cream cheese to a bowl and mix them together. Make sure that everything is well combined.
4. Brush the mushrooms with your melted butter. Next, spoon the filling into each of the mushrooms.
5. Lay some parchment out onto a baking sheet and place the mushrooms out on it. Sprinkle the tops with your favorite grated cheese.
6. Bake them for about 20 minutes. Broil them to crisp up the cheese and enjoy.

Kale, Lambs Quarters, and Cheese Manicotti

You'll Need:

- Parsley
- Pepper and salt
- Grated parmesan, .5 c
- Mozzarella, 1.5 c – divided
- Beaten eggs, 2
- Ricotta cheese, 18 oz
- Finely chopped kale, 1 c
- Finely chopped lamb's quarters, 1 c
- Spaghetti sauce, 2 to 3 c
- Cooked manicotti shells, 1 box

You'll Do:

1. Start by mixing together the kale, lamb's quarters, ricotta, egg, parmesan, pepper, salt, and 1 cup of mozzarella.
2. Carefully fill each of the cooked manicotti shells with the filling.

3. In a casserole dish, evenly spread a layer of spaghetti sauce over the bottom. Lay the stuffed manicotti on top of the sauce. Make sure that you don't lay one manicotti on top of another.

4. Spread the rest of the spaghetti sauce over the shells and then sprinkle on the rest of the mozzarella cheese.

5. Once your oven is at 350, bake for 30 minutes.

6. Allow this to cool for about five minutes and then serve. Garnish with some parsley if desired.

Purslane Egg Cups

You'll Need:

- Favorite spices
- Cheddar cheese, .25 c
- Eggs, 12
- Milk, .25 c
- Finely chopped onions, 2
- Small pepper, chopped
- Chopped purslane, 2 c

You'll Do:

1. Start by making sure your oven is at 350. Take a muffin pan and grease it well.
2. In a pan with some butter, add the pepper and onion and sauté for about five minutes.
3. Using a food processor or blender, add the spices of your choosing, milk, eggs, and cheese. One the pepper and onions are cooked, add those in as well. Blend everything together.
4. Pour into a bowl and stir in the purslane and any other wild greens you would like to use.
5. Divide the egg mixture between the cups in the muffin pan and bake for 20 to 25 minutes, or until the eggs are completely cooked.

Stuffed Milkweed Pods

You'll Need:

- Bread crumbs
- Boiled and split milkweed pots, 20
- Pepper and salt
- Chopped jalapeno
- Cooked bacon, 2 slices
- Diced red onion, 1 tbsp
- Softened cream cheese, 4 oz

You'll Do:

1. Start by setting your oven to 375.
2. Place the cream cheese in a bowl with the pepper, salt, bacon, jalapeno, and onion. Mix everything together until well combined.
3. Go through the boiled milkweed pods and get rid of any immature seeds and silk from them. Spoon in about two teaspoons of the cream cheese mixture, filling the pod until full.
4. Roll the seam of filling into the bread crumbs and then sit them seam side up on a parchment-lined baking sheet.
5. Bake them for 15 to 20 minutes.

Weed Burgers

You'll Need:

- Favorite spices, 1 tbsp
- Finely chopped wild edibles of choice, 1 c
- Breadcrumbs, .75 c
- Finely chopped onion
- Chopped garlic, 3 cloves
- Salt, .5 tsp
- Beaten eggs, 4
- Cooked quinoa, 2.66 c

You'll Do:

1. Start by mixing together the spices, salt, eggs, and quinoa. Mix in the garlic and onion. Lastly, stir in the wild edibles and breadcrumbs. Allow this mixture to rest for a few minutes so that the breadcrumbs can absorb the moisture.
2. Form into patties and fry them in a pan for about five minutes on each side.
3. You can enjoy the burgers on buns just like you would a regular burger.

Wild Potato Pancakes

You'll Need:

- Pepper, 1 tsp
- Salt, 1 tsp
- Garlic powder, 2 tsp
- Flour, 2 tbsp
- Hemp seeds, 2 tbsp
- Chopped mushrooms, .25 c
- Chopped wild greens, .5 c
- Chopped onions, .33 c
- Eggs, 2
- Grated potatoes, 4 c

You'll Do:

1. Start by melting some butter in a pan. Add the onions and sauté for a couple of minutes and then mix in the mushrooms. Cook for another two minutes and set it off of the heat.

2. In a bowl, add the spices, flour, seeds, eggs, sautéed vegetables, and grated potatoes. Mix everything together for about three minutes to make sure everything is well combined.

3. In a frying pan, melt a bit of butter to make sure the pancakes don't stick. With your hands, create balls from the batter. As you do this, you will notice that the potatoes contain some liquid. Squeeze out some of this liquid.

4. Lay the ball onto the frying pan and flatten it with a spatula to about a half-inch thick. Cook until both sides are golden brown. Add extra butter if you need to so that they don't stick.

5. Continue until all of the batter is used and enjoy.

Wild Roasted Cabbage

You'll Need:

- Crumbled cooked bacon, 5 slices – if desired
- Grated cheese – if desired
- Pepper and salt
- Ground dried stinging nettle, to taste
- Chopped garlic, to taste
- Chopped onion, to taste
- Chopped garlic mustard, to taste
- Olive oil, 4 to 6 tbsp
- Head of cabbage

You'll Do:

1. Start by getting your oven to 350.
2. Lay the cabbage on a chopping board and slice into ¼ inch thick slices. Do your best to hold the leaves together. You can usually get four to six slices.
3. Brush the bottom of the slices with oil and then lay them on a parchment-lined baking sheet.
4. Add the rest of the oil to a bowl and combine with the garlic, pepper, salt, and nettle. Brush this on top of the cabbage slices.

5. Sprinkle the tops of the cabbage with the onions and garlic mustard.

6. Bake this for 20 minutes. Take out and top with the cheese and bacon if you are using them. Let this back for about 15 minutes more. Enjoy.

Buttered Chickweed

You'll Need:

- Pepper
- Salt
- Butter
- Finely chopped onion
- Chopped chickweed, 2 c

You'll Do:

1. Start by washing the chickweed. Bring a pot of salted water to a boil and add the chickweed into the water. Let this cook for a couple of minutes and then drain well.
2. Add some butter to a pan. Add in the onion and sauté until translucent. Add in the chickweed. Season with some pepper and salt or any other spices that you would like.

Plantain Salad

You'll Need:

- Salt, 1 tsp
- Wine vinegar, 1/8 c
- Olive oil, 1/8 c
- Chopped garlic, 1 to 2 cloves
- Finely chopped celery stalk
- Thinly chopped onion
- Can of drained chickpeas
- Finely chopped cabbage, .5 c
- Chopped plantain, 2 c

You'll Do:

1. Start by mixing all of the above ingredients together, except for the vinegar and oil. Place the in the refrigerator. Once this has chilled well, add in the vinegar and oil.
2. If you find that the salad is a bit on the dry side, you can add some more vinegar and olive oil. Taste and adjust any seasonings that you need to.

Blueberry Labrador Tea

You'll Need:

- Lemon juice, 1 tbsp – if desired
- Water, 1 c
- Blueberries, 1.5 c
- Labrador tea, 4 c

You'll Do:

1. Add the blueberries and water to a pot and let them come to a boil. Turn the heat down and simmer, stirring often, until the blueberries start to break down. This will take about five to ten minutes. Add the blueberries into the brewed tea. Add in your favorite sweetener, stir, and then let it come to room temperature.
2. Place in the fridge until cold, usually about two hours. Strain the mixture into a pitcher and mix in the lemon juice.

Burdock Tonic Tea

You'll Need:

- Dried peppermint leaves, to taste
- Dried red clover flowers, 2
- Dried dandelion root, 1 tsp
- Dried burdock root, 1 tsp

You'll Do:

1. Start by mixing all of the ingredients together and then place them into a large mug.
2. Pour in some boiling water, cover the cup, and let it steep for 30 minutes. Strain and enjoy.

Healthy Heart Tea

You'll Need:

- Cold water, 5 c
- Fennel seeds, 2 pinches
- Ginger root, 2 slices
- Hawthorn berries, 1/8 c
- Motherwort, 1 tsp

You'll Do:

1. Pour the water into a pot and add in all of the herbs. Let this come to a boil, turn the heat down, and simmer for 20 minutes.
2. Let it cool enough to drink, and add in lemon or honey if you want. Enjoy.

Highbush Cranberry Juice

You'll Need:

- Maple syrup, 1 tsp
- Orange juice, 2 c
- Highbush cranberries, 2 c
- Water, 3 c

You'll Do:

1. Start by adding the water to a pot and letting it come up to a boil. Turn off the heat. Add the berries to the water and mash them a bit. Allow this to sit for 30 minutes.
2. Strain the water and allow the liquid to cool. Once cool, add in the syrup and orange juice. This will keep in the fridge for five days.

Immune Boosting Coffee

You'll Need:

- Pinch of salt
- Water, 4 c
- Coffee, .5 c
- Powdered turkey tails, 1 tbsp
- Powdered tinder conk, 1 tbsp
- Powdered chaga, 1 tbsp

You'll Do:

1. Start by adding the water to your drip coffee maker. Place the turkey tail, tinder conk, and chaga into the filter and then add the coffee on top of them. Add in a pinch of salt to help with the bitterness.
2. Brew as you normally work, and sweeten if desired.

Fennel and Angelica Cookies

You'll Need:

- Flour, 2.5 c
- Fennel seeds, 1 tbsp
- Chopped angelica leaves, 2 tbsp
- Light beaten egg yolk
- Sugar, .5 to .75 c
- Butter, 1 c

You'll Do:

1. Start by adding the butter and sugar to a bowl and mix well. Stir in the angelica and egg yolk. Slowly add in the flour and fennel. Stir everything together so that it is well combined.
2. Once blended, cover, and refrigeration for 30 minutes.
3. Get your oven to 375 and place some parchment paper on some baking sheets.
4. Take the dough out and roll golf ball sizes of dough between your hands. Flatten them out on the parchment to eighth of an inch thick. You can also roll the dough out and use cookie cutters if you would prefer.
5. Bake for 12 to 15 minutes. Let them cool for ten minutes and then place them on a wire rack to cool completely.

Bee Balm Cookies

You'll Need:

- Orange zest, 4 tsp
- Chopped bee balm flowers and leaves, 4 to 5 tbsp
- Flour, 1 c
- Powdered sugar, .5 c
- Butter, .5 c

You'll Do:

1. Start by beating the butter with the bee balm, sugar, and orange zest until well combined. Add in the flour and mix together. You may have to use your hands to really get it mixed together because it will get thick. Make sure you don't overwork the dough once you add the flour. Once smooth, roll into a cylinder and wrap in parchment. Chill for two hours.
2. Once chilled, slice the dough into quarter-inch slices. Place them on a baking sheet about an inch apart.
3. Bake for eight to ten minutes at 350. Enjoy.

Coltsfoot Sorbet

You'll Need:

- Coltsfood flowers, .5 c
- Sugar, .25 c
- Water, 2 c

You'll Do:

1. Start by adding the water and sugar to a pot. Stir until the sugar has dissolved.
2. Add in the coltsfoot and let it come up to a boil. Turn the heat down and cook for five minutes. Set this off of the heat and let it cool.
3. Pour into your ice cream maker and fix it according to the directions. You can enjoy this now or place it in the freezer for later.

Dandelion Banana Bread

You'll Need:

- Baking soda, .5 tsp
- Baking powder, 1 tsp
- Dandelion flower petals, .33 c
- Flour, 1.25 c
- Brown sugar, .33 c
- Egg
- Olive oil, .5 c
- Ripe banana

You'll Do:

1. Start by mashing up the banana and mixing in the sugar, egg, and oil. Stir in the baking soda, baking powder, dandelion flowers, and flour. Mix until everything comes together. If you want, you can also add chocolate chips or walnuts.
2. Scoop into a greased loaf pan.
3. Bake for 20 to 25 minutes at 350.
4. Once cooked through, slice and enjoy.

Honey Cattail Cookies

You'll Need:

- Coconut butter, 4 tbsp
- Honey, .5 c
- Cinnamon, 3 tsp
- Vanilla, 1 tsp
- Oats, 2 c
- Coconut flakes, 2 c
- A brown cattail

You'll Do:

1. Start by grinding the cattail fluff up with the oats to create fluffy flour. Add this to a bowl along with the remaining ingredients. Knead until thoroughly mixed and then form into cookies.
2. You can eat this raw, or you can place them in the oven for one to two hours at the lowest setting. Enjoy.

Nutty Plantain Snack

You'll Need:

- Sea salt
- Olive oil
- Sesame seeds, 3 handfuls
- Pumpkin seeds, 3 handfuls
- Plantain seeds, 1 handful

You'll Do:

1. Start by adding all of the seeds into a bowl. Add in the salt and olive oil, making sure that the seeds are coated well in the oil. You can also add in other spices if you would like to.
2. Spread this out on a baking sheet and roast for ten to 15 minutes at 250.
3. Let them cool and enjoy.

Pine Cookies

You'll Need:

- Vanilla, 1 tsp
- Eggs, 3
- Melted butter, .5 c
- Red or white pine powder, 8 tbsp
- Sugar, 1.5 c
- Flour, 3 c

You'll Do:

1. Start by adding the dry ingredients into a bowl.
2. In a separate bowl, mix together the vanilla, eggs, and butter. Stir the wet ingredients into the dry ones until well combined.
3. Roll the dough into balls and lay them on a cookie sheet. Using a fork, press the cookies out to a quarter-inch thick.
4. Bake them for ten to 12 minutes at 325. Enjoy.

Pine Rum Balls

You'll Need:

- Shredded coconut, 2 tbsp
- Ground up pine needles, 2 tbsp
- Rum, 2 tbsp
- Half of a pine cake, crumbled

You'll Do:

1. Start by breaking up the pine cake into small pieces. Add in the rum, slowly, until you can start forming the cake into balls. If you need to, you can add in a bit more rum.
2. In a different bowl, combine the coconut and pine.
3. Roll the cake balls into the coconut and pin mixture. You can also use powdered sugar and chocolate sprinkles if you would prefer.
4. Place them on a plate to chill in the fridge for two hours. Serve and enjoy.

CONCLUSION

Thank you for making it through to the end of the book, let's hope it was informative and able to provide you with all of the tools you need to achieve your goals whatever they may be.

The next step is to start exploring the world of edible wild plants. While it may sound like a hard thing to start doing, foraging for your own food can be a rewarding experience. You don't have to replace your entire diet with foraged foods. Just trying some of the edible wild foods will make a difference, and you will start to realize it's not as hard as it may have originally seemed.

The most important thing to remember is to be nice to the planet when you are foraging. Don't just ravage an area because you find a lot of good food. Take only what you are going to use and make sure that the plants will be able to replenish themselves before you return to that spot again. Be ethical in your foraging.

Also, never go foraging blind, especially while you are still new at it. Make sure you have something with you that can help you identify plants. Better yet, take a trip through an area that you would like to forage in and just study the plants to figure out what you've got. Once you have identified some safe plants, you can come back at a later time and forage. The last thing you want to do is assume that something is safe because it "kind of looks like it" and then get sick, or worse, once you eat it.

Lastly, remember that a lot of these edible wild plants can also be used to treat various ailments. Our ancestors only had plants to cure health problems. It's even believed that for every disease mankind suffers from, there is a plant that can cure it. Unfortunately, modern medicine has made it seem like plants can't help treat ailments. I'm not telling you that herbs and plants should be used in place of modern medicine, it shouldn't, but you can tap into the power of plants from time to time to help supplement your doctor's treatment. If you feel like you have a serious ailment, always seek a doctor's opinion first.

Above all else, enjoy the foraging process. Commune with nature and learn something from her.

Finally, if you found this book useful in any way, a review on Amazon is always appreciated!